161 558192 5

D0245008

TRACING YOUR AIR FORCE ANCESTORS

TRACING YOUR AIR FORCE ANCESTORS

Phil Tomaselli

Pen & Sword
FAMILY HISTORY

First published in Great Britain in 2007 by
PEN & SWORD FAMILY HISTORY
an imprint of
Pen & Sword Books Ltd
47 Church Street
Barnsley
South Yorkshire
S70 2AS

ISBN 978 1 84415 573 6

A CIP catalogue record for this book is available from the British Library.

Typeset in Palatino and Optima by
Phoenix Typesetting, Auldgirth, Dumfriesshire

Printed and bound in England by
CPI UK

Pen & Sword Books Ltd incorporates the Imprints of
Pen & Sword Aviation, Pen & Sword Maritime, Pen & Sword Military, Wharncliffe Local History, Pen & Sword Select, Pen & Sword Military Classics and Leo Cooper.

For a complete list of Pen & Sword titles please contact
PEN & SWORD BOOKS LIMITED
47 Church Street, Barnsley, South Yorkshire, S70 2AS, England
E-mail: enquiries@pen-and-sword.co.uk
Website: www.pen-and-sword.co.uk

CONTENTS

Introduction		1
Chapter 1	The Beginnings	8
Chapter 2	The Royal Flying Corps in the First World War	18
Chapter 3	The Royal Naval Air Service in the First World War	50
Chapter 4	The Creation of the RAF	65
Chapter 5	The RAF between the Wars	73
Chapter 6	The Second World War: Main RAF Commands	89
Chapter 7	Other records for the Second World War	103
Chapter 8	The RAF after 1945	127
Chapter 9	Medals, Casualties and Courts Martial	139
Appendix 1	Ranks in the RFC/RNAS/RAF	159
Appendix 2	Royal Flying Corps – Order of Battle	161
Appendix 3	RNAS Order of Battle, 31 March 1918	164
Appendix 4	RAF Order of Battle in France, 11 November 1918	167
Appendix 5	Fighter Command Order of Battle, July 1940 (The Battle of Britain)	169
Appendix 6	Bomber Command Order of Battle, May 1942	173
Appendix 7	AIR Papers at the PRO which may Contain Information Useful to Family Historians	177
Appendix 8	Useful Books	181
Appendix 9	Other Useful Archives, Collections and Sources	186
Appendix 10	Magazines and Periodicals	192
Appendix 11	Websites	193
Appendix 12	RAF Abbreviations	196
Index		209

ACKNOWLEDGEMENTS

First and foremost I owe my wife, Francine, the warmest and most heartfelt thanks for living with this interest for longer than we've been married and for putting up with my regular visits to TNA and other archives. I also owe a large debt to her late father, Peter Pepper, whose own knowledge of aircraft dwarfed my own, and who was a most useful mentor, guide and inspiration.

The staff at The National Archives have always been helpful, courteous, kind and knowledgeable. Simon Fowler, previously editor of *Family History Monthly* and now *Ancestors*, adopted me early on and has encouraged me (and acted as editor) ever since. Rosemary Horrell loaned me, and allowed me to copy, the papers of her late stepfather, Group Captain Reggie Bone, in the first place and got me started on the whole project.

Swindon and Marlborough Library staff have been patient in searching out obscure books for me through the inter-library loan service. Jan Keohane, Catherine Rounsfell and Duncan Black at the Fleet Air Arm Museum allowed me access to the extensive archives there and offered me advice, assistance and some good laughs and gossip while I was there. Martin Kender shared with me his researches into Lt Guy. Various friends in the Western Front Association and Cross and Cockade International have shared their interests and enthusiasms with me; David List, Paul Baillie and William Spencer and other researchers at TNA have given me advice, help and support over many years. Richard Davies and his assistants at Leeds University Library helped me with the Liddle Archive and kindly provided the photograph of C P O Bartlett. Thanks to the staff of the RAF Museum, Imperial War Museum and Museum of Army Flying; Nick and Carmen Coombs for information on, and the photograph of, Nick's grandfather Reginald Heath; Joyce Cooke for the photographs of the late Ron Cooke and for permission to use his log book.

Various veterans of the First World War and later who I've been in touch with deserve special mention, particularly Guy Blampied, William Savage and John McGowan. Now all passed away but each, in their own way, was keen to help me in various aspects of my researches.

My father Phil Tomaselli senior did his National Service in the RAF in the early 1960s and has encouraged me and talked me through his service record (as well as producing his own history of RAF Yatesbury to act as a spur). Finally, I'd like to dedicate this book to my late mother, Doreen Tomaselli, who from an early age said that I should be a writer and researcher and who lived to see my first articles published. She would have been delighted to see my name on this book.

INTRODUCTION

At the height of Second World War some 110,000 officers and 1,050,000 other ranks were serving in the Royal Air Force, with many thousands of others in related air services such as the Fleet Air Arm, Glider Pilot Regiment and the Army's Air Observation Posts. Given that there were some 300,000 men and women in the Royal Air Force at its formation on 1 April 1918 and that the origins of British military aviation go back to the 1870s, very many of us have ancestors who flew for their country or supported the men who did.

My own interest in the RAF began in the 1980s when I was given the papers of Group Captain Reginald Bone, who'd served with the Royal Naval Air Service and then the RAF (with a short gap, when he was Director of Civil Aviation in Egypt) until 1941. There wasn't then (or if there was I couldn't find it) a general guide to RAF records, and service records from the First World War had not been released, so researching what I hoped to turn into Reggie's biography was challenging to someone whose main interest, previously, had been the Victorian army. But it was fun . . . and the interest in the RAF and in service records and archives has stayed with me ever since. I hope this book will provide information about how to find information about individuals but also give pointers and examples to researchers (particularly family historians) about where they can find records of units and bases individuals served in, the kind of work they did, the aeroplanes they worked with and the people they worked alongside.

Family history, if it is to mean anything, means taking individuals and understanding their world. History isn't just created by kings, politicians and air marshals but by the men and women who actually did the fighting in the air, the servicing of the aircraft on the ground, who drove the fuel bowsers, moved the aircraft symbols on the plotting tables, manned the telephone exchanges and typed the orders. Without them and their sacrifices of time, effort and sometimes health or lives, no great events would be possible. I hope this book will help you fit your ancestor into the events in which they participated.

Inevitably, when seeking an individual, you'll find some records have been destroyed, mislaid, not been released or have been so indexed that their title gives no hint that useful information is contained inside. Persevere, think laterally, try looking further up the command chain for copies of documents that might be relevant, seek out other

Wing Commander R J Bone RAF, probably taken in late 1918 following his transfer from the RNAS to the RAF. (Mrs R J Bone via Mrs R Horrell)

45

NAMES IN FULL Bone Reginald John CB CBE DSO

Date and Place of Birth 2 October 1888 Father's Name and Address

(per birth certificate) Dorking, Surrey

Dates of Commissions and Substantive Promotions. P.=Permanent. S.S.=Short Service.			Rewards and Distinctions, including "Mentions in Despatches."		Qualifications, Examinations and Completed Courses (including types of machines flown and sub-classifications).	Special attainments, Languages and Athletics.
Date of Gazette.	Rank.	Date of Effect.	Reward.	Date of Gazette.		
[illegible]	[illegible]	17.10.13	[illegible]	Not gazetted	[illegible]	Ripley Court Surrey 1898–1902
[illegible]	Fl Lt RNAS	1.7.14	[illegible]	–	[illegible] Hy S+S (26)	Northwood Park Winchester 1902–[illegible]
	Flt Cdr	23.[illegible].15	D.S.O.	[illegible]	[illegible]	HMS Britannia [illegible] cadet training ship [illegible] 1907–08
	Sq [illegible]	30.6.16			[illegible] 2/18 – 5/18 (D.260)	Power [illegible]
[illegible]	Major RAF	1.4.18	[illegible]		[illegible] Southampton [illegible] (May 28)	
[illegible]	Lt Cdr [illegible]	[illegible]	C.B.E. (Mil)	[illegible]	Attended instrument flying [illegible] Capable instrument flying pilot	Rugby, soccer, hockey, golf (D.260)
[illegible]	Maj (Sq Ldr) [illegible]	[illegible]	[illegible]	[illegible]		Riding, fencing (ACR'24)
[illegible]	[illegible]	[illegible]	[illegible]		Passed [illegible] flying [illegible]	Colloquial French (ACR'24)
[illegible]	Wg Cdr RAF	1.[illegible]	C.B. (Mil)	1.1.34 (4)		
1.7.29 [illegible]	Group Captain	1.7.29				
[illegible]	Retires at own request	15.4.34				
			1914–15 Star; B.W. & V. Medals	–	[illegible] Southampton, Fairey [illegible] B.F. (ACR'27) [illegible]	

√ Previous Service. Medical Boards.

Disability.	Date of Commencement.	Date and Place of Board.	Category.	Date of Next Board.	Remarks.
[illegible]		[illegible]		[illegible]	[illegible]

Excerpt from Reggie Bone's service record showing his RNAS and RAF service prior to 1934. (Via Mrs R Horrell)

Date of retirement 2.10.41 [illegible]

√ Special Reports, Commendations, etc.

Date of Gazette	Non Substantive Appointment	Date of Effect
[illegible]	[illegible]	[illegible]
[illegible]	[illegible]	[illegible]
[illegible]	[illegible]	[illegible]
[illegible]	Placed on half pay list Scale A	15.10.33

[illegible] Admiralty for meritorious work in connection with air attack on [illegible] [illegible] Feb 11th, 12th & 14th 1918 [illegible] "Good leader [illegible] command." [illegible] "a most suitable officer [illegible] command [illegible]" [illegible] "Has commanded [illegible] with great efficiency. Very fine pilot." [illegible]

174395/32. To be placed on half pay, S.A. on 15.10.33 – pending retirement.

RECOMMENDED FOR PROMOTION

By			Date
AOC. C. [illegible]	1/2	A	Dec. 1934
[illegible]	6/6	A	June 1935
AOC. India	1/4	A	June 1929

Neurasthenia. Slightly [illegible] Slight concussion abrasions etc. 23.2.29. 260606/33.

2.11.17. 3.11.17. Re-survey 14.12.17. Re-survey 19.4.18. Central Hosp. 1.6.21. Central Hosp. 26.5.21. Central Hosp. 20.9.21. Central Hosp. 8.11.24 Portland Harbour (C.Y. 1072). 6.5.24 A1B. 12.2.26 A1B. 9.13.98 [illegible] A1B.

Excerpt from Reggie Bone's service record showing commendations and opinions of senior officers. (Via Mrs R Horrell)

R.A.F. OFFICERS RECORD OF SERVICE
CONFIDENTIAL

AIR MINISTRY (R.A.F. OFFICERS APPOINTMENTS AND ... OFFICE) — A.M. FORM 1406

SURNAME BONE — CHRISTIAN NAME(S) REGINALD JOHN — PERSONAL NUMBER 01045

C.B., C.B.E., D.S.O.

DATE OF BIRTH 2nd October 1888 — PLACE Dorking Surrey — RELIGION C of E

(2) DATE AND TYPE OF COMMISSION & SUBSTANTIVE PROMOTIONS: RANK	DATE	LONDON GAZETTE
For details of Commission See old Record Sheet		
Retires at own request	15/ 4/34	24/ 4/34
Re-employed on Active List	4/ 9/39	87(a) List 63
Reverted to Retired List	14/ 1/41	A142318/40

(3) NON-SUBSTANTIVE PROMOTIONS: For details of Actg Ranks See Old Record Sheet

GROUP CAPTAIN R. J. BONE

DECEASED

Officer Died 29.8.72.
Notification – Times 31.8.72.

Excerpt from Reggie Bone's service record showing his Second World War service and notice of death. (Via Mrs R Horrell)

collections. The simple act of researching can be a lot of fun in its own right! Whatever you find, please write it up – this piece of advice comes from someone with half a dozen completed research projects that just require writing up, so I know it's difficult, but please set it down for the benefit of other family and other researchers who might come after you.

I've thoroughly enjoyed the researching and writing of this book, which has taken me into areas I'd not previously looked at, such as the early Royal Engineer balloon units and the RAF during the Suez Crisis of 1956. I hope this book will encourage you to explore new areas too.

I had very little knowledge of aviation or the RAF when I started and am assuming the same level of knowledge in my readership, though I hope there will be material here for the more experienced researcher too.

The structure of the book

I've tried to keep the structure pretty much chronological. I've also, where possible, separated the book into sections devoted to the different services that have contributed to the development of British military aviation. As you read on, however, it will become apparent that it was perfectly possible for an officer or man to enlist in one service (the Army or Royal Navy) and then be transferred to the Royal Air Force, or even to join one part of the Army (the Royal Engineers) then be transferred to another (the Royal Flying Corps) then into the Royal Air Force. Some men seem to have managed to serve in just about every service at one time or another, though on the whole these are rare. It certainly can be confusing for the family historian and it means that frequently more than one set of records has to be consulted just to get a complete service record, let alone to build up a complete picture of what an ancestor did.

The book begins with the pioneers of military aviation, the Royal Engineers, who flew balloons during Queen Victoria's reign; by the reign of Edward VII they were already exploring manned kites and dirigible (steerable) airships and were experimenting with powered aircraft. In 1912 the Royal Flying Corps was established from the Royal Engineers and a group of pioneer naval pilots – this union lasted until 1914 when the Royal Navy created its own Royal Naval Air Service and left the Royal Flying Corps to the Army. The two services were amalgamated into the Royal Air Force in 1918. In 1936 the Admiralty succeeded in winning back control of naval aviation and created the Fleet Air Arm, and at about the same time the Army began its own experiments flying light aircraft to spot for the artillery. During the Second World War the Army also created the Glider Pilot Regiment and, during the 1950s merged these with the artillery spotters to form the Army Air Corps. All of these various units and services will be looked at.

I've finished the book with the end of National Service in 1962, but records of units and bases, and some relating to individuals, such as court martial papers, have been released post-1962 (and continue to be released under the thirty-year rule). The basic procedures for finding these remain the same: squadron records are still in AIR 27 for example, so information can be found for men and women who served as Regulars after that date.

Types of records

There are two main types of official records for the air services, which will be useful to the family historian. Personal records give information about individuals, their date of birth, dates of enlistment and discharge, next of kin, sometimes a physical description, and details of where they served. Unit records will tell you what the squadron, or station, the man (or woman) was posted to was doing in the period that they served with them and help you to flesh out the bones of their service history. Some service records are already open to the public, others will need to be applied for. Unit records from the earliest days up until the 1970s are generally open. We'll look at both in some detail and explain how to apply for records that remain closed.

There are a huge number of technical and policy records which are also open to the public in the various AIR files at TNA which you might want to explore for background. I'll mention these from time to time but it's not really within the scope of this book to go into detail.

Getting started

Ancestors who served in the Royal Air Force and its predecessors, the Royal Flying Corps, Royal Naval Air Service and Royal Engineers Balloon Section are easier to trace than ancestors in the other services. Though there are gaps caused by administrative errors or by losses in wartime these are as nothing compared to the irreparable damage done, for example, to the army records from the First World War by German bombs during the Second World War. The diligent researcher should be able to find their ancestor's service record and, using this, find records of where they served and what they did.

Experienced family historians will already be aware that the more you know about an ancestor at the start of your quest, the more you are likely to find out as you go on. Collect together all the information you can find and write it down.

Gathering information

Look out any paperwork you might have; speak to your relatives, particularly the older ones, to see if they have any medals, log books, official papers, letters or memorabilia that might be useful to you. Ask if they remember any stories they might have been told about where your relative served and what they did – though such stories need to be treated with a certain amount of caution there is often an element of truth in them.

USEFUL THINGS TO HAVE AT THE START

Full name of the person you're researching and date of birth

Name(s) of likely next of kin (father, mother, spouse)

Likely address at time of enlistment

Their service number if you have it (on discharge papers or round the rim of a First World War medal)

Their rank, if you know it

Some of this information you'll need in order to obtain a copy of the individual's service record or to check that you've found the correct service record if it's already been released to the public.

Any photographs, letters, postcards, pieces of uniform or other kit that come from their period of service can help illustrate your research and may give further clues to exactly what they did and where they went.

Personal service record

Everyone who served in the RAF or any of its predecessors or associated flying services (such as the Fleet Air Arm or Army Air Corps) has a unique service record, which gives basic details of what they did during their service, where they went, their promotions, medals and next of kin. We'll look in more detail at service records later, but the first thing to decide is whether their records might have been released and be available at the National Archives, or whether you'll need to apply to the Ministry of Defence for a copy.

Service records for officers who served in the Royal Air Force, Royal Flying Corps and Royal Naval Air Service and left before the end of 1919, as well as for other ranks who served and left before the end of about 1922 are publicly available at The National Archives at Kew in London. There are no plans as yet to put these on-line so a visit will be required if you want a copy, or you can employ a professional researcher to get them for you.

Service records for both officers and men who served after these dates remain closed and are held by the appropriate section of the Ministry of Defence. For the

Royal Air Force the records are at: PMA (Sec) IM 1b, Room 5, Building 248a, RAF Innsworth, Gloucester, GL3 1EZ (tel. 01452 712612, ext: 7622).

If you are seeking information on a living relative you'll need to get them to complete a 'Subject Access Request Form' which can be downloaded from the Veterans Agency website at http://www.veteransagency.mod.uk/service_records/service_records.html. There is no charge for veterans or their widows/widowers to receive this information.

If the veteran and spouse are not surviving you will need either to be their next of kin, or have their next of kin's written authority to have information released to you. RAF Innsworth will provide you with a Next of Kin form to complete which will explain who the next of kin are (all children count equally, for example). It may occasionally be necessary to provide proof of identity. There is a search fee (2007) of £30.00.

Other things you can do

While you wait for the service record there are a few things you can do. There are some very good histories of the RAF (see Appendix 8) which your local library may have, or which they can obtain for you through the inter-library loan service. Try looking at the RAF's own website http://www.raf.mod.uk which will tell you about the current role and capabilities of the service, but also contains a great deal of information about its history and ethos. There are other excellent websites devoted to aspects of the RAF or to individuals which you can look at. I particularly recommend that you start looking at the website for The National Archives.

Another website I wholeheartedly recommend for demonstrating quite how far you can go with personal research and what you can do with it is at: http://myweb.tiscali.co.uk/mikeskeetsww2website/index.html. This site is dedicated to the author's father and is a fascinating and poignant personal history, as well as a most useful source on some of the areas looked at in this book.

Reading an individual's service record

Though the names of the actual forms used to record an individual's service may have changed, the details on the forms and the information recorded remain pretty constant. Personal details and details of next of kin will be familiar to family historians, but the purely military information can be a little baffling.

The most important details concern Movements, i.e. the units he or she was sent to, and Promotions. Movements are the hardest to read as all too often the units are given as initials or abbreviations. There is a list of some of the most popular abbreviations and initials from both wars in Appendix 12, as well as details as to how to find more.

Service records for First World War and Second World War also detail 'Casualties, Wounds, Campaigns, Medals, Clasps, Decorations, Mentions etc' This should give brief details of any wounds, the medals the person was entitled to and to any Mentions in Despatches. It will, occasionally, give additional details of their postings.

There should be at least some information regarding Character and Trade Proficiency, which will tell you a little about their service. Unfortunately for many early RAF service records very little is recorded, mainly because they are derived from records originally compiled by the Army or Navy. It is always worth looking

– Pay Clerk Robert Percy Smith joined the RFC in June 1917 and obviously enjoyed his work as he extended his service in 1919. After the formation of the RAF in April 1918 he seems to have received regular reviews of both his character and trade efficiency, his work consistently being described as 'satisfactory' and his character as 'V G' (Very Good). In spite of some of the escapades my father has told me about from his days doing National Service I'm pleased to say that his conduct was described as 'Ex' (excellent)!

If your relative was subject to any court martial or discipline resulting in loss of time towards pay or pension you'll find brief details in the section 'Time Forefeited', which will give number of days forfeited and usually a reason. If the offence was serious enough to warrant a court martial then an offence and date of court martial should be recorded.

The service record for Sidney Dilworth, an RFC Fitter, shows that he lost two days pay on 8/9 August 1917 because he was first in civil custody then held in the Guard Room. He was tried by DCM (District Court Martial) and sentenced to 85 days detention for absence. In November 1917 he was sentenced to 48 hours detention and loss of 28 days pay for failing to report on discharge from hospital, and in March 1918 he lost another 6 days pay for absence. For all his problems over absence he seems to have been a competent Aircraftsman – he was wounded in the foot in October 1914 and received the 1914 Star, showing that he went out to France in the first months of the war. He was also temporarily promoted to Acting Corporal Mechanic in September 1918, though this was cancelled in December the same year.

How accurate is the service record

This is a question I have occasionally been asked, and it appears to be the case that not everything is recorded. My father is adamant that he attended a Potential Officer's Assessment Course at RAF Biggin Hill shortly after being called up for his National Service in 1959. He is quite sure that he got special treatment from his Senior NCO during his square-bashing training, just in case he was accepted and came back as an officer to make the man's life a misery later. There is no record of this on what remains of his record.

Reggie Bone, the RNAS/RAF officer who first interested me in the Air Services, was equally clear in his unpublished autobiography that he was in Russia during 1918, but again there is no evidence of this on his record. I also know of a man whose service record moves him straight from the UK to Gallipoli, but who was torpedoed en route (from other evidence) – there is no mention of the incident on the record.

The thing to remember is that the service records are usually compilations derived from various sources accrued during the period of service and that they only record incidents and movements that will affect a man's pay, pension or medal entitlement. Some small movements or temporary postings may not have been felt to be relevant. Treat the service record as the bare bones of a career – if there is other evidence which suggests an unrecorded gap or posting follow it up by looking for unit records or other material.

Chapter 1

THE BEGINNINGS

The Royal Engineers – fathers of modern military flying

The Montgolfier brothers made the first balloon flights at the end of the eighteenth century and balloons were talked about as weapons during the Napoleonic Wars, but it was really only by the 1860s that the technology was developed that allowed balloons to be used for observation.

During the American Civil War both sides used balloons for directing gunfire and spotting their opponents' movements. During the Franco-Prussian War of 1871, sixty-five balloons were used to lift 164 passengers – including the French Minister Léon Gambetta – and 9,000 kg (20,000 lbs) of mail from the besieged city of Paris by flying over the German lines. In Britain, the authorities began to sit up and take notice.

In 1878 the Royal Engineers made a series of experiments with balloons at Woolwich Arsenal and a Balloon Equipment Store was created. A few officers began to make regular flights and Balloon Sections accompanied expeditions to Bechuanaland in 1884/5 and to Suakin on the coast of the Red Sea in 1885. In 1890 the Balloon Section, Royal Engineers, was created and a School of Ballooning was formed in 1892 at Laffan's Plain, Aldershot. During the Anglo-Boer War of 1899–1902 three Balloon Sections were sent to South Africa and proved invaluable in providing maps drawn from aerial observation, in directing gunfire and spotting Boer positions. It has been said that the military balloons were one of the few bright spots in a campaign better known for its errors and failures.

In 1903 Colonel John Edward Capper RE, an officer with distinguished service in South Africa behind him, was appointed Commandant of the Balloon School. That same year the Wright brothers made their first flight at Kitty Hawk. Capper was ideally placed, with contacts both official and unofficial in the world of aviation, to hear of their success and understand what it meant. The Royal Engineers Balloon Sections were soon to acquire some extra responsibilities.

Victorian soldiers' records

Officers' records: engineer officers

Tracing officers is always easier than tracing ordinary soldiers (officers are not only more important, but they also tend to write the records!). There are two published lists for the period available on the open shelves of the Microfilm Reading Room at

TNA. Large regional libraries may have copies of the official War Office list or be able to obtain copies of either list through the inter-library loan system at modest cost.

The official *Army List* was published quarterly by the War Office and has an index of officers, which will show you the page(s) your officer appears on. You can trace his previous promotions and his later career by looking at earlier and later editions. The list also contains a list of retirements, casualties and officers who have ceased to serve for other reasons (such as dismissal, cashiering and transportation for various military offences) but, though it gives the name and regiment, it rarely gives any more information.

Towards the end of the nineteenth century the *List* introduced a section, separate from the main listing, with officers' war services, telling you about which campaigns they served in and occasionally giving details of individual actions.

Another source of information is the privately published *Hart's Army List* which is invaluable for most of the nineteenth century, though by the time of the Boer War the official War Office list carried much of the same information. It is always worth checking as *Hart's* does occasionally contain additional material. Copies of *Hart's* are also on the shelves in the Microfilm Room at TNA.

Regiments kept records of officers' service and for the Victorian Royal Engineers these can be found in WO 25/3913–20, which are on microfilm at TNA. The records are more or less chronological by date of joining the Corps, though there is sometimes a degree of overlap so more than one volume may have to be consulted. Fortunately each volume is indexed at the start, so this shouldn't take long. These records will give you basic information about the officer's enlistment, where he served and when, which actions he took part in and his family (if married).

Ronald Joseph Henry Laws Mackenzie's record is in WO 25/3915. He was born on 15 April 1863 and entered the Royal Engineers through the Cadet Company on 25 July 1882. He is recorded as '*Suakin Expedition 1885*: Employed from 6th March to 8th May with Balloon Detachment RE at Suakin'. He was also at the Battle of Hasheen and as acted as Adjutant at RE Base and commanded the Telegraphs and Signalling sections at Suakin. He later served as Assistant Survey Officer in the Zhob Field Force in 1890 and the First and Second Miranzai Expeditions in 1891 as Survey Officer. He seems to have suffered from occasional periods of ill health and went on half pay in December 1903 for this reason and retired on a pension of £200 per year on 8 August 1906.

Don't forget that many relatively junior officers would have continued to serve into the First World War or even beyond. If they rose to high rank their papers may not yet have been released if they continued to serve after 1922 but many will feature in *Who's Who* in the years after the war and this will give you some idea of their service. You will need to contact the Army Records office at Glasgow to request their papers.

Other ranks

WO 97 is the place to start when looking for a Royal Engineer Other Ranks' service records. This series contains enlistment and discharge papers for men who survived their army service and were discharged. Discharge papers for 1873–82 are between WO 97/1849 and WO 97/1857 (for Royal Engineers only), for 1883–1900 are between WO 97/2172 and WO 97 4231 and for 1900–1913 they are between references WO 97/4232 and WO 97/6323. For the two later periods they're collected alphabetically for the whole army. Some misfiled papers have been gathered together and are in alphabetic sequence between WO 97/6355 and WO 97/6383. The records you can see are the originals and come in boxes you will need to look through to find the record you want. They're generally in order but you'll usually need more information on your relative than just a name as even unusual names duplicate – place of birth is generally useful as this is given on the top right-hand side of each document. Please be considerate to other researchers when searching each box and do not muddle up the papers.

Individual soldiers spent only part of their career working with balloons and then returned to the Corps as a whole. This was to spread knowledge of balloons and ballooning and to create a tactical reserve of men who could be called upon in time of war if the balloon section needed to expand.

The records will tell when and where a man enlisted, provide a physical description, list promotions and next of kin, and where a man served and for how long. There should also be a note giving the reason for a man's discharge. Please note that the papers are unlikely to specify a connection with ballooning, but if you know your man was awarded a medal as part of a balloon contingent you should find his records here. Unfortunately not all records seem to have survived.

One thing that you will have to watch out for is the number of men who served in the Boer War who were still serving, or rejoined the colours, at the start of First World War. Most soldiers joined up for a short period of active service followed by a few years in the reserves, but a fair proportion, who decided early on to make a career of the army, would have enlisted for twenty-one years. Many soldiers, after the end of their active service, decided to re-enlist rather than go back to civvy street. A good number also chose to re-enlist voluntarily at the start of the First World War and were asked if they'd already served, in which case their papers would have been amalgamated with their First World War records. If they went back into the Royal Flying Corps then check the records held in AIR 79 but if only back into the Royal Engineers you'll need to check the Soldiers' papers in WO 363 and WO 364.

Unfortunately, if your ancestor went back into the Royal Engineers, about 60 per cent of First World War soldiers' records were destroyed in the Blitz, but if you can't find anything in WO97 try checking the Medal Cards for the First World War, available on-line at TNA website or on microfiche at Kew. If your ancestor served abroad in the First World War, and was thus entitled to a medal, you can at least have a good idea that his papers were amalgamated. TNA holds, on microfilm, the remaining papers (some of them only partial because of fire damage) of First World War soldiers and their website will tell you how to go about looking for these.

Samuel Burness: a Victorian Royal Engineer and balloon man

Though it doesn't say so in his service records, we know that Samuel served with balloons because his name appears in the list of men awarded the Egyptian Medal with clasp for Suakin for the RE Balloon Section in 1885 (WO 10/64 ff80). His service record is in WO 97/2411. The covering paper is a Short Service Attestation for one year dated 6 June 1900 and inside this is his original Attestation from 30 September 1874. Presumably he enlisted again for a short period in 1900 when the Boer War meant that most of the Regular Army were serving in South Africa and emergency measures were taken to keep the Home Forces up to strength.

It is the papers that accompany his 1874 Attestation that will be of most interest

Excerpt from the service record of Royal Engineer Samuel Burness showing his service in Egypt and the Sudan

to a family historian. They show that he was 15 years old when enlisting, born in Australia (the 1900 Attestation says Melbourne) and give his trade as 'Musician'. He enlisted for twelve years (plus an additional twelve months if at war at the time of discharge). He was 5' 1¾" tall with a fair complexion, blue eyes, brown hair and with a wart on the front of his neck. After service as a Boy Soldier he was posted to the ranks as a Sapper on 1 September 1877 and was rapidly promoted to Lance Corporal (1 June 1880), Second Corporal (30 September 1882), Corporal (15 February 1883) and Sergeant (1 November 1885). He served in Bermuda for nearly three years between 1878 and 1881 and married Elizabeth Oakes 'without leave' (i.e. without his Colonel's permission) at Frinsbury-Stroud (Kent) on 28 August 1882. He served in Egypt and the Sudan between 16 February and 7 July 1885, qualifying for the Egypt Medal with Suakin Clasp and also the Khedive's Bronze Star. After further Home Service he again went to Bermuda in October 1886, having re-engaged for a further nine years so as to be eligible for a pension, but deserted on 4 November 1886.

It's possible to make an informed guess why. With his marriage made 'without leave' he would have had to have left his wife behind in England. On 19 July 1886 she'd given birth to their first child, Albert Sydney, and may have been struggling to cope with her husband abroad. He rejoined the Corps on 5 July 1888 and was tried for desertion and reduced to the ranks. He was again promoted and reached the rank of Corporal on 23 June 1892 but deserted once more in August 1893. His wife had had a daughter, Daisy, on 13 May 1889 and another son, Edmund Matchett on 23 July 1893. Perhaps Samuel's desertion was connected to this? He rejoined in September 1893 but this time was charged not only with desertion but with embezzlement, forgery and loss of kit. Once again he was reduced to the ranks, but this time he served nine months' imprisonment. He returned to the Corps on 23 July 1894 and continued to serve as a Sapper until he was discharged as a result of reaching the end of his second period of engagement on 20 June 1899. He can't have been a bad soldier, in spite of his crimes, as apart from his periods actually deserted or in prison he was allowed almost all his service to count towards his pension, and they seem to have been keen to take him back in 1900.

The papers also record the birth of his final child, Nellie, on 16 June 1896 and note that, at some point, his wife was brought onto the strength of the Corps (which meant she received a small amount of pay for laundry duties and would be looked after in the event of his death). His medical papers show he was treated for syphilis in 1877 and for bronchitis in 1888. He was a typical Victorian soldier, though without perhaps the addiction to drink that plagued so many of them.

Balloon unit records

Most of the early balloon unit records were passed to the Air Historical Branch when they began compiling the 'Official History of the War in the Air' so are held in the AIR 1 class at TNA. The most important run of files is between AIR 1/728/176/3 and AIR 1/732/176/6/148 which are papers donated by the Royal Aircraft Establishment. There, among papers dealing with dirigibles, gliders, kites and early aeroplanes you'll also find, among many others:

'Names of balloons serving in South Africa, China and Australia 1884–5': AIR 1/728/176/3/1.

'Work done by 3 Balloon section RE in South Africa 1900–1901': AIR 1/728/176/3/3 .

'Lecture on the role of English military balloons in the South African War by Colonel Lynch': AIR 1/728/176/3/8.

'Lecture on the role of English military balloons in the South African War by Major Trollope': AIR 1/728/176/3/9

'Report on Balloon Section at manoeuvres, 1903': AIR 1/728/176/3/10.

'Report on Balloon Section at summer training, 1903': AIR 1/728/176/3/11.

'Report on work of 1Balloon Section at Army manoeuvres, 1904': AIR 1/728/176/3/19.

There are reminiscences of early balloons 1884–5 in AIR 1/2310/220/1, notes on the technical training of Balloon Sections in AIR 1/1608/204/85/36 and a report on balloon operations at Gibraltar in 1905 in AIR 1/1611/204/87/23.

Some material seems to have been missed by the Air Historical Section and can still be found in the War Office (WO) series, including:

'Employment of Major Templer as military balloonist to Royal Engineer

Committee. His inventions and salary. Reports on experiments. 1884–7': WO 32/8584.

'Pay and promotion of officers, arms and equipment, pay and conditions Of enlistment for the Balloon Corps, 1886': WO 33/46.

'Report by Major H Elsdale, Royal Engineers, on the comparable position of British and Foreign Ballooning, 1886–6': WO 32/6067.

'Proposed Balloon Depot and Training Establishment at Chatham, 1886': WO 32/6068.

Reports on operations

The Balloon Sections took part in only four campaigns, so records are correspondingly few and far between. Some of the ones given below will have to be searched for references to the Balloon Sections.

For the Suakin Campaign of 1885 (that Samuel Burness took part in) there are papers in WO 32/6134 (Report on the Royal Engineers at Suakin), WO 33/44 (diary of the Suakin Expedition), WO 28/374 (diaries of the Nile and Suakin Expeditions) and WO 106/223 (diary of the principal events at Suakin).

WO 32/6134 records a flight by Lieutenant Mackenzie (who we met in the 'Officers records' section) on 25 March 1885. The Balloon Detachment accompanied a convoy, moving in the form of a square, guarding against ambush from any direction by an enemy that was considerably swifter and capable of ambush at almost any point. The balloon was inflated in advance and fastened to a wagon that accompanied the convoy. Lieutenant Mackenzie was sent aloft with his instruments and kept 'at altitudes varying from 200 feet to 400 feet according to his wishes'. He spent seven hours in the air before being replaced by Sapper Wright, who flew for another two hours. Mackenzie and Wright observed and reported on the movement of British troops in the vicinity and of small parties of the enemy. It was noted that 'The aeronaut could see small bodies of the enemy at five miles distance' and that the square was not delayed by the accompanying balloon detachment. Mackenzie was able to report 'Small bodies of enemy to our left front 800 yards off' and to reply to the question 'What strength is enemy' with the answer 'About 40 men and 30 men'. Presumably one of the new telephones was used to link the balloon and the ground.

The Bechuanaland Expedition of 1885 (which saw no fighting) is covered by WO 32/8204 (Report of Officer Commanding Royal Engineers) with additional material in WO 33/44 (Report of Proceedings of the Bechuanaland Field Force 1884–5), WO 33/45 (Correspondence relative to the Bechuanaland Field Force 1884–5), WO 106/263 (Quartermaster General's Diary of the Bechuanaland Field Force) and WO 147/35 (Report on the Bechuanaland Expedition).

The Boxer Rebellion (Relief of Peking) of 1900–1 is covered by WO 32/6059 (Report on French Balloon Section at Peking and employment of 4th Balloon Section, Royal Engineers in China) and WO 106/76 (Final Report on the engineer operations of the British contingent).

The work of the Royal Engineers generally in the Boer War (1899–1902) is thoroughly covered in a series of reports, WO 108/283-298, which in the main cover different areas of the country and phases of the campaign.

Heavier than air machines

On 17 December 1903, Wilbur and Orville Wright succeeded in achieving the dream that went back to the legend of Icarus. Over the course of the day they made four powered and controlled flights in a heavier than air machine.

The Wrights began negotiations with the British Government to sell the patents and the working machine to them. Over the next three years negotiations dragged on, the principal British negotiator being Colonel Capper, with the British feeling that the Wright brothers' conditions were too strict (and hoping that a British inventor could come up with an all-British machine). In 1906 a Brazilian, Santos Dumont, built and flew his own machine. In January 1908 an Englishman, Henry

Farman, flying a French aircraft, made a circular flight of 1 km and secured a prize of 50,000 francs.

The first man to fly in England, and perhaps the most famous man in the early days of aviation in Britain, was an American, adopted as honorary Briton, 'Colonel' Samuel Franklin Cody (real name Franklin Samuel Cowdery). He'd worked as a trail cowboy, sharp shooter and horseman in a touring circus and an actor and changed his name so that people would associate him with 'Buffalo Bill' Cody. He had become interested in kites and had begun the development of a man-carrying kite for the Royal Engineers. These allowed a series of kites to carry an observer hundreds of feet into the sky to spot for artillery and were considered quite a success.

Using his experience building kites and gliders, Cody was able to build 'Army Aeroplane No. 1', a large and clumsy experimental machine no doubt based in part on information that Capper had gained from examining the Wright machine. It was Cody who made the first proper controlled flight of 1,390 feet on Laffan's Plain, Farnborough, on 16 October 1908. Cody also worked closely with Capper and the Royal Engineers on developing an army airship, which they flew together over central London. Other officers learned to fly privately and were seconded to the Royal Engineers for experiments in long-distance flying and reconnaissance.

The formation of the Royal Flying Corps

On 1 April 1911 the Air Battalion of the Royal Engineers was formed but its existence was to be short lived. In April 1912 a White Paper set out the planned future of military and naval aviation in Britain. Both fledgling aviation sections were to be combined into 'The Royal Flying Corps' which was to have naval and military wings. A Central Flying School (CFS) was established at Upavon in

An early private flying school, Eastbourne Aviation Company c.1912. Note the variety of aircraft. Most early aviators learned to fly at schools like this. (Author's collection)

205th T. D. S.

Nort Flight.

25 June 1918.

REPORT ON ACCIDENT.

NAME OF PILOT Charles Arthur Matthews

NO. of MACHINE Avro 6412 Type ______ NO. of ENGINE ______ Type ______

INJURIES TO PILOT OR PASSENGER (if any) Slight bruise on thigh

PROBABLE CAUSE and details of ACCIDENT Insufficient petrol therefore not enough power

LOCALITY OF ACCIDENT Aerodrome.

CONDITION OF MACHINE Badly damaged
(State here "Slightly Damaged", "Badly Damaged" or "Wrecked").

PILOT'S REPORT.

I was unable to get the correct mixture on the fine adjustment; the engine then failing to give sufficient revs, lifted a few feet off the ground & swung round and ~~dropped~~ went into the ground.

Signature C. A. Matthews

Countersigned S. B. Joyce.
Flight Commander.

This accident was due to sheer carelessness on behalf of the Pilot. S.B.J.

An RFC crash record from 1 Training Depot School. (Author's collection)

Wiltshire which was commanded by a naval officer but whose adjutant was a 40-year-old army officer who had just passed his brevet, Hugh Trenchard.

The CFS was not expected to train pilots in the basics of flying. This was left to private tutors at the many flying schools then springing up. Any officer who successfully passed his Royal Aero Club brevet (the pilot's licence required before you could call yourself an aviator) privately would be given £75 towards the cost of his training. Naturally, the price of basic training at the flying schools soon settled at £75, which encouraged tutors to get pupils through the course and test as quickly as possible. The government saved money though, as only officers who passed had their training paid for and those who failed (or crashed and were injured or killed) did not feature as an expense.

Once having passed their brevet the officer would receive his £75 and shortly be posted to Upavon for training in the military aspects of flight. Here he was trained in long-distance flying, recognition of ships and military formations from the air and technical aspects of engines, ailerons, rudders and joysticks. After the end of this training he'd be appointed to a squadron or station. By July 1914 there were 259 officers serving, or under training, in the combined wings of the RFC, 102 in the naval wing and 157 in the military (plus 52 in the reserves).

On 7 August 1913 Colonel Cody took off with a passenger for a short flight in his new hydroplane from Farnborough. At about 500 feet the aircraft folded up in mid-air and pitched groundwards. A soldier reached the scene and found the bodies of Cody and his passenger lying together, still and quite dead. Flying Corps officers led away his stepson who had rushed to the scene. A later investigation concluded that if both men had been wearing seatbelts they might have stayed with the plane and survived. As befitted his position as designer of the first military aircraft, and in honour of his association with the Royal Engineers and Farnborough, he was given a military funeral at the Military Cemetery at Thorn Hill, Aldershot. The Royal Flying Corps made all the arrangements and thousands turned out to pay their respects.

Records on the early days of the RFC and Army aviation are in AIR 1 and include: AIR 1/669/17/122/792 Experiments with 'Cody's' kites 1903 to 1909; AIR 1/728/176/3/29 Proposed experiments with flying machines, 1905; AIR 1/729/176/4 Early experiments with dirigibles, gliders and aeroplanes 1906–9;

AIR 1/729/176/4/2 Royal Aircraft Establishment Reports on various experiments and tests 1906–1907. The report on Cody's death is in AIR 1/823/204/5/63.

The early Royal Naval Air Service

The Admiralty were also interested in the business of flying and had themselves employed Colonel Cody to experiment with man-carrying kites. With the increasing range of ships' guns it was felt necessary to have the ability to see over the horizon and there was also the perceived threat of German naval Zeppelins which might give the enemy an advantage in reconnaissance during any battle. Unfortunately the Royal Navy had their fingers burnt when an airship they'd commissioned, 'The Mayfly', lived up to its name and broke its back before its first flight. This made the Admiralty reluctant to commit time and money to new technology that might become another expensive failure. A wealthy businessman and pioneer aviator, Frank McClean, offered two of his own aircraft; another aviator promised to teach flying at no cost and Short Brothers, who had an aircraft factory at Eastchurch in Kent, offered technical support.

It was an offer the Admiralty could not refuse and, from over 200 applicants, they selected four young officers for initial training. Lieutenants Samson, Longmore, Gregory RN and Gerrard RMLI. (Royal Marine Light Infantry) began their six-month training course at Eastchurch on 1 March 1911. On 25 April Samson and Longmore passed their Royal Aero Club brevet and a few days later, on 2 May, Gregory and Gerrard joined them.

On 10 January 1912 sailors carefully shipped a Short biplane out to HMS *Africa*, a battleship which had an improvised flight deck built of planking. The aircraft, piloted by Samson, ran down the flight deck and soared into the air. Samson had proved that aircraft could take off from moving ships, though it was to be some years before the RNAS worked out quite how to get them down safely onto the ship again!

Samson and his brother officers trained other officers and some ratings to fly and continued to study aircraft types and flying techniques.

Records on early naval aviation before the First World War include: AIR 1/2400/293/1 History of Eastchurch Air Station, Isle of Sheppey 1909–1926; AIR 1/2467 Aviation course at Eastchurch: work done by trainees and future training; ideas for aeroplanes in naval warfare; AIR 1/2469 Aviation course at Eastchurch: weekly reports, syllabus etc; AIR 1/2462 RNAS progress report 1912; AIR 1/2463 Differences between RFC and RNAS 1912; AIR 1/2444 Accident to HM Airship No.1: Court of Enquiry finding 1911; AIR 1/2438 Proposals for development of RNAS 1913, AIR 1/651/17/122/451 Training of naval officers in aviation, 1911.

The split between the RFC and RNAS

For three years the two wings of the Royal Flying Corps (naval and military) co-operated quite happily and officers from both services attended the Central Flying School at Upavon. Reggie Bone, a young naval officer who transferred from the Submarine Service, attended the 4th Course at Upavon 1913/14 and wrote:

> The diversity of mess kits at dinner was extraordinary for there were sappers, gunners, cavalry and line – and No. two infantry officers came from the same

regiment. Thus with some naval mess kits the mess was an unforgettable picture at dinner time.

Naval officers took part in Army manoeuvres and there was much exchange of technical information between the two wings. Unfortunately the Admiralty did not feel that the union was satisfactory and in July 1914 separated the naval wing to become The Royal Naval Air Service.

The Army and Navy Lists

Because (at least in theory) the RFC and RNAS were both wings of the Royal Flying Corps, some details of both appear in the Army and Navy Lists. The Army List gives few details of the RFC apart from listing the officers and giving their seniority. Because it is indexed it is possible to follow a man's progress as he is promoted, but details about postings are not given. The Navy List, however, is much more detailed about its own particular wing but just gives basic details about the officers of the RFC. The Royal Navy List for January 1915, for example, details all the staff and instructors at the Central Flying School at Upavon (it was a joint responsibility of both wings and had staff from each) as well as all the pupils, both Army and Navy. It also lists all the RNAS officers and the stations and ships at which they were based, as well as pupils under training at private flying schools. For the RFC the Navy List gives only the officer's rank and seniority.

The RN List continues to give listings of postings for individuals right the way through to the March 1918 edition, so it is possible to trace an officer's career this way, which can be important if he didn't transfer to the RAF or if his service continued beyond 1919 (in which case his service record is not yet released).

Chapter 2

THE ROYAL FLYING CORPS IN THE FIRST WORLD WAR

The RFC went to France with four squadrons, sixty-three aircraft and approximately 900 men. At home a rapid training programme was established to produce pilots, observers and ground crew who for the first two years of the war were largely sent to France to build up an establishment of squadrons, now grouped together in wings (see Appendix 2) so that there were twelve squadrons by September 1915 and twenty-seven squadrons, with over 600 aircraft, by the start of the Battle of the Somme in July 1916.

At first aircraft were poorly armed (if at all) but gradually they were mounted with light machine guns (usually a Lewis gun) or, in two-seater aircraft, a larger machine gun to be fired by the observer. The Dutch inventor, Anton Fokker, developed a means of firing a machine gun through the propeller and sold his invention to the Germans so that during the summer and autumn of 1915 the RFC suffered heavy casualties. Eventually a British mechanism was developed that allowed British machines to do the same and the aircraft gradually became specialized as fighter, bombers or reconnaissance aircraft, and the squadrons specialized accordingly. Developments in engines, cameras, clothing and weaponry meant that the aircraft of 1918 were technologically much more advanced than the ones originally sent to France in 1914.

By mid-1916 the RFC was able to take back responsibility from the RNAS for the air defence of Britain. On 2 September 1916 Lieutenant William Leefe-Robinson of 39 (Home Defence) Squadron shot down a Zeppelin north of London. The airship crashed in a spectacular ball of flame that lit up the night sky for miles around, boosting public morale after several Zeppelin raids had taken place, apparently with impunity. Leefe-Robinson was awarded a Victoria Cross.

As well as their squadrons in France the RFC sent men and aircraft to the Middle East, Italy, and the Balkans, with special missions also being sent to Russia to train their flying corps, and to Canada for recruitment and training.

The Germans themselves developed new aircraft so that it was a constant race between the opposing flying services, with periods when one or the other was in the ascendancy. At times a new pilot's life expectancy on being sent to the Western Front could be measured in days. Gradually new training techniques, and the posting of new pilots to home defence duties to start with, meant that casualties reduced.

Having gone out to France in 1915 from his position as Adjutant at the Central

Flying School, Hugh Trenchard rose rapidly to command the RFC in France. He pushed his pilots hard, but visited them regularly and took careful notes of their problems and complaints. He was much respected by his men. In 1918, after arguments with Lord Rothermere, the Secretary of State for Air, he resigned as head of the RAF in France and was appointed to command the Independent Force of the RAF, a force of heavy bombers to attack Germany. In 1919 he was appointed as Chief of Air Staff and remained in the role until his retirement in 1929.

The Royal Flying Corps service records

Officers

The place to start if you're looking for an officer who served with the RFC is in the AIR 76 records at TNA. These record the services of the 26,000 officers of the RAF who left the service before the end of 1919 and are held on microfilm at TNA,

Christian Names (in full) John William
Surname Patterson
Date of Birth 31 Aug. 1894
Permanent Home Address 46 Broughton Rd, Handsworth.
Present Service Engagement
Name of Person to be Informed of Casualties: Name O.W. Patterson; Address 46 Broughton Rd, Handsworth. B'ham.
Relationship Father
Date Sheet Started 17 APR 1918
Registered Papers No.

MOVEMENTS: Outgoing Authority | Unit from which | Unit to which | Date of Effect | Special Remarks re Duties | Incoming Authority Confirming

MOVEMENTS (continued): Outgoing Authority | Unit from which | Unit to which | Date of Effect | Special Remarks re Duties | Incoming Authority Confirming

APPOINTMENTS AND PROMOTIONS: Date of Gazette | Rank | Description | Graded as | Date of Effect

Excerpt from J W Patterson's service record showing his First World War service.

Lt John William Patterson RAF, Santa Lucia, Piave, Italy, 1918. (With thanks to the Patterson family)

arranged alphabetically. Even though they are RAF records they include details of the officers who transferred from the RFC and of many of the officers who were killed before 1 April 1918.

As a basic guide they give the surname, usually the Christian name, name and address of next of kin, home address, dates of appointment to units served in, appointments and promotions, details of any gallantry medal entitlement and short resumes of any medical boards. If the officer had served previously with another unit (and most RFC officers had served with a regiment before transferring to the RFC) this is usually mentioned. If the officer had a special skill or qualification this is often mentioned, as is his occupation in civilian life. Officers who had previously served in the RNAS may have a separate sheet stamped 'Naval' with brief details of their RNAS career.

Trevor Frederick George Strubell was born on 28 October 1885 and joined the RFC as a 2nd Lieutenant and Assistant Equipment Officer in June 1916. Previously he is shown as having served with the Hertfordshire Yeomanry in Egypt and qualified as a machine gunner. He was married, and his wife (recorded only as Mrs Strubell) lived at 105 Cromwell Road, South Kensington. Starting as an Assistant Equipment Officer he soon rose to the rank of Major and was appointed as Commanding Officer of the RFC Aircraft Park in Canada in January 1917. He returned to Britain in July 1918 and served as a Staff Officer 2nd Class at the Air Ministry and with No. 8 Group RAF. He had only one medical board, on 24 April 1918 and passed 'A'. He was transferred to the Unemployed List on 28 May 1919.

His rapid rise through the ranks to the rank of Major may be explained by the special qualifications he is noted as having:

> 10 years experience steam and petrol engineering gained in shops and running all business organisations of oil mill and subsequently entire mechanical transport and coffee grinding machinery of Brazilian Warrant Co Ltd, Santos. Brazil. Experience of Curtis and Depardussin aeroplanes (pre war). Has flown both types.

His civilian occupations are given as engineer and seed crusher at E Manser and Co. of Hertfordshire, Herts., from 1903 to 1912, then employed by Brazilian Warrant Co. Ltd, Santos, Brazil from January 1913 to 3 August 1914. It is probable that he resigned his job at the outbreak of the war, returned home and joined the local Hertfordshire Yeomanry. Given his wide knowledge of machinery, the fact that he'd flown aircraft before the war and had a proven track record as a manager, the RFC must have snapped him up when he applied to transfer to them!

War Office records

The RFC was part of the Army so most of its officers began their career with a regiment and then transferred. This means that there are often records of their previous army service, along with some details of their RFC service, in the officers' personnel files from the War Office. These are held at TNA in their WO 339 and WO 374 series. Very many officers' records were destroyed during the Blitz in the Second World War

The WO 339 series contains 139,908 records and correspondence for Regular Army and Emergency Reserve officers who served in the First World War. The WO 374 series contains nearly 78,000 service records for officers given either a Territorial Army commission or a temporary commission. Both series can be searched using the surname but be aware that with a common surname this might take some time. TNA website and paper guides at Kew give hints which might help to speed up your search.

The content of the files varies enormously, from a note supplying date of death, to a file of several parts containing attestation papers, record of service, personal correspondence and various other information.

Records of British reserve officers who were commissioned into the Indian Army were originally held separately, but later added to this series.

The AIR 76 record for John William Patterson, a pilot with 34 Squadron in Italy at the very end of the war gives only limited information about his service before he joined the RFC in 1917. His service record however survives in WO 374/52684 and, despite having been weeded in 1955 of information the War Office no longer considered relevant (it says on the cover), it gives a wealth of information about him and his previous service. He was born at Aston in Birmingham on 21 August 1894, was son of O W Patterson, a commercial traveller, and was himself, at the time of his enlistment an export merchant's clerk. He enlisted at Recruiting Office No. 3 at the Curzon Hall, Birmingham, where, after some delay, he was enlisted as a clerk in the Ordnance Corps with the Regimental No. 08761. After completing his basic training at Woolwich he was posted to Egypt in July 1916, where he served until March 1917.

2nd Lt J W Patterson, 6th King's Liverpool Regiment prior to his attachment to the Royal Flying Corps in 1917.

B Flight 34 Squadron, Italy, November 1918. (With thanks to the Patterson family)

At this point, being described as 'thoroughly sober', intelligent and well educated, he was recommended for a commission as an infantry officer. After an interview at the War Office he was posted to No. 21 Officer Cadet Battalion at Crookham. In his application for a commission he stated, in order of preference, the Ordnance Corps, the RFC and finally, the Infantry. He was not sufficiently technically qualified for the Ordnance Corps but was, instead, posted as a 2nd Lieutenant to the 6th Battalion Liverpool Regiment. There is nothing on the file about his transfer into the RFC and subsequently the RAF, though it is noted that he was placed on the Retired List on 8 March 1919. The file also has a copy of his will and various medical boards he went through.

Other officers' records

In the records of some RFC squadrons you can find both 'Joining Certificates' and 'Particulars of Officers Forms' which may be of use to the family historian. The Joining Certificates for 34 Squadron are in AIR 1/1396/204/26/19 and give a wealth of information including full name, date of birth, original regiment (as noted, most officers originally enlisted with another unit), next of kin, place(s) of education, profession, religion, previous service, dates of joining the army, the RFC and the squadron, date of obtaining 'wings' and types of machine flown. Similar information appears on the 'Particulars of Officers' forms, which should also include the squadron flight that the officer was attached to. From a combination of the two forms we can see that William Robert Bathurst Annesley, a former pupil of Cheltenham College and St Andrew's University, became a Royal Engineers officer on 24 September 1914. He was attached to the RFC on 26 July 1915 and saw service as an Observer at Gallipoli. He became a Flying Officer on 26 December 1917, having flown a dozen different aircraft including the modern SE5 and RE8. He was 26 years and 10 months old in March 1918 when he joined 34 Squadron and his next of kin was his mother, Mrs E M Annesley. His civilian trade was as a mechanical engineer.

Royal Flying Corps – other ranks' service records

The Royal Flying Corps was part of the Army service and I have seen it argued that the records for its other ranks (non-commissioned officers down to private soldiers) were retained by the Army after the war. This does not seem to be the case.

The RFC service records, at least in the form of each man's attestation papers, were passed to the Air Ministry who then, in most cases, transferred the information onto the personnel record sheets that form the basis of TNA's AIR 79 collection.

Official No. 403 CHRISTIAN NAMES (in full) Arthur Edward

1 BIRTH Date

Place: Town Birmingham; Parish St George's; County Warwick

2 RELIGIOUS DENOMINATION C of E

3 CURRENT ENGAGEMENT IN H.M. FORCES:—

(a) State whether in Army, Navy or R.A.F. Army

(b) Period:—

	Date Current Engagement Commenced	Age at that Date	Duration of Period of Engagement
Boy Service			
Man Service	24 9 12	20 5/12	11 C. 4 R.

(c) Date of actual entry into R.F.C. 24 9 12; R.N.A.S.; R.A.F.

(d) Period engagement extended for

(e) Period re-engaged for

4 PRIOR ENGAGEMENT IN H.M. FORCES:—

Unit	Period From	Period To	Rank on Discharge

5 CIVILIAN OCCUPATION

(a) Description Carpenter

(b) Name and Address of last Employer

MOVEMENTS * Att = At

Departure Authority	Unit from which	Unit to which
		6 A.R.S.
6 W.D. 137	6 A.R.S.	53 T.D.S.
6 W.D. 169	53 T.D.S.	50 T.D.S.
	50 T.D.S.	120 Sqdn
	120 Sqdn	5 A. Ac Pk
		CENSUS
	8 AA Pk.	S.P. Henley

Emb'k F. 2. Sq. 1. A.P. Home Est.

An RFC service record for Arthur Edward Hobson (AIR 79/6)

This seems even to have applied to men who had been killed, died of wounds or had been released from service prior to 1 April 1918, so that AIR 79 should be the place to start if looking for any RFC other ranks.

The records are held in bound volumes organized on a numerical basis so you'll need to know your ancestor's RFC service number before you can locate the record. If you have his medals the number will be round the rim, along with his name and rank. You can then use the number to locate the relevant volume.

If you don't know the service number then you'll need to consult the index to airmen and airwomen's service records which is held in TNA's AIR 78 series, on microfilm in the Microfilm Reading Room. This series covers all airmen and women between 1918 and 1975, so for First World War people the number you're looking for will be below 329,000. Each film records a given range of names (AIR 78/1 covers 'Aagenson, Elizabeth to Agent, Walter', for example, and there are four whole films of 'Smith'). Some films are also not the easiest to read, but with patience you should be able to find the service number to lead you to the correct volume of AIR 79.

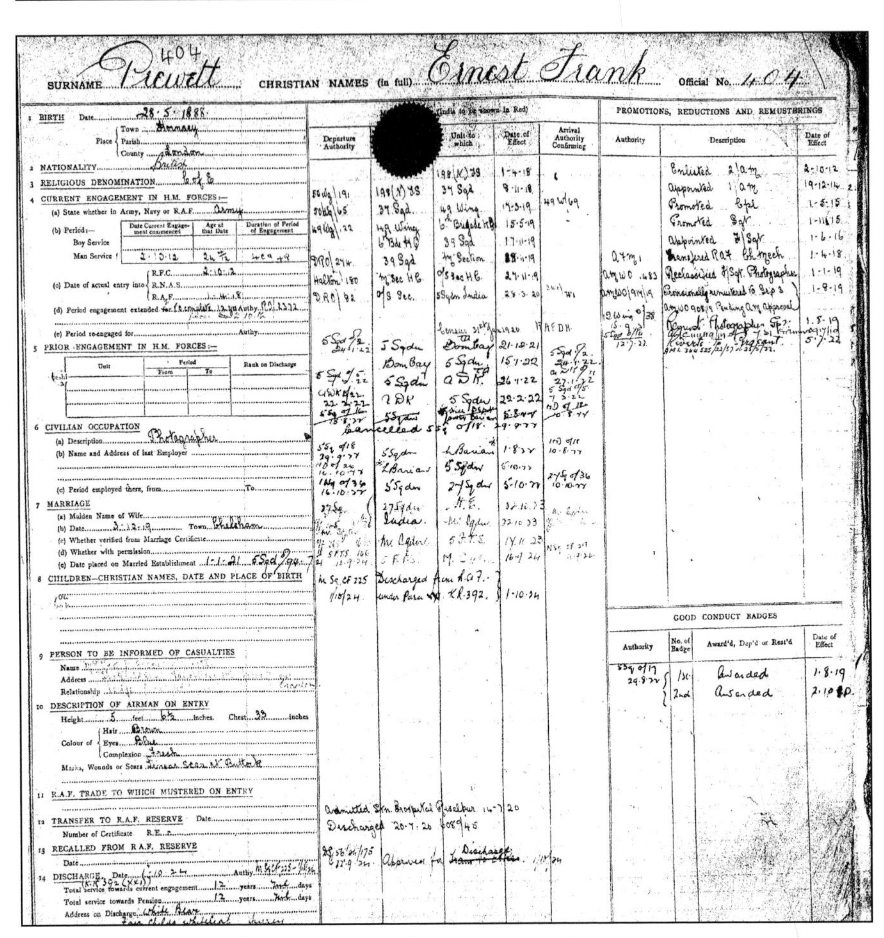

SURNAME Drewett 404 CHRISTIAN NAMES (in full) Ernest Frank Official No. 404

1 BIRTH Date 28.5.1888
Place: Town Hornsey; Parish; County London
2 NATIONALITY British
3 RELIGIOUS DENOMINATION C of E
4 CURRENT ENGAGEMENT IN H.M. FORCES:—
(a) State whether in Army, Navy or R.A.F. Army
(b) Period:— Date Current Engagement commenced / Age at that Date / Duration of Period of Engagement
Boy Service
Man Service 2.10.12 / 24 7/12 / 4 ea + 8
(c) Date of actual entry into R.F.C. 2.10.12; R.N.A.S.; R.A.F. 1.4.18
(d) Period engagement extended
(e) Period re-engaged for ... Authy.
5 PRIOR ENGAGEMENT IN H.M. FORCES:— Unit / Period From / To / Rank on Discharge
6 CIVILIAN OCCUPATION
(a) Description Photographer
(b) Name and Address of last Employer
(c) Period employed there, from ... To
7 MARRIAGE
(a) Maiden Name of Wife
(b) Date 3.12.19 Town Cheltenham
(c) Whether verified from Marriage Certificate
(d) Whether with permission
(e) Date placed on Married Establishment 1.1.21
8 CHILDREN—CHRISTIAN NAMES, DATE AND PLACE OF BIRTH
9 PERSON TO BE INFORMED OF CASUALTIES
Name
Address
Relationship
10 DESCRIPTION OF AIRMAN ON ENTRY
Height 5 feet 6½ inches. Chest 33 inches
Colour of Hair Brown; Eyes Blue; Complexion Fresh
Marks, Wounds or Scars
11 R.A.F. TRADE TO WHICH MUSTERED ON ENTRY
12 TRANSFER TO R.A.F. RESERVE Date
Number of Certificate R.E.
13 RECALLED FROM R.A.F. RESERVE
Date
14 DISCHARGE Date 1.10.24 Authy.
Total service towards current engagement 12 years days
Total service towards Pension 12 years days
Address on Discharge

Departure Authority	Unit from which	Unit to which	Date of Effect	Arrival Authority Confirming
		198 (K) TS	1-4-18	
	198 (K) TS	39 Sqd	8-11-18	
	39 Sqd	49 Wing	17-3-19	49 W 69
	49 Wing	6th Bde HQ	15-5-19	
	6 Bde HQ	39 Sqd	17-11-19	
	39 Sqd	Ph. Section	18-11-19	
	Ph. Sec HQ	Ph Sec HQ	27-11-19	
	Ph Sec	5 Sqdn India	28-3-20	
	5 Sqdn	Bombay	21-12-21	
	Bombay	5 Sqdn	15-1-22	
	5 Sqdn	ADK	26-1-22	
	ADK	5 Sqdn	22-2-22	
	5 Sqdn		5-8-22	
	5 Sqdn	Bannu	1-8-22	
	Bannu	5 Sqdn	5-10-22	
	5 Sqdn	27 Sqdn	5-10-22	
	27 Sqdn India			
			17-11-23	
			16-9-24	
	Discharged from R.A.F. under Para KR. 392.		1-10-24	

Admitted ... Hospital ... 14-7-20
Discharged 20-7-20

PROMOTIONS, REDUCTIONS AND REMUSTERINGS

Authority	Description	Date of Effect
	Enlisted 2/AM	2-10-12
	Appointed 1/AM	19-12-14
	Promoted Cpl	1-5-15
	Promoted Sgt	1-11-15
	Appointed F/Sgt	1-6-16
	Transferred RAF Ch. Mech	1-4-18
	Reclassified F/Sgt. Photographer	1-1-19
	Provisionally remustered ...	1-8-19
	Remustered Photographer ...	1-8-19
	... Sergeant	5-7-22

GOOD CONDUCT BADGES

Authority	No. of Badge	Award'd, Dep'd or Rest'd	Date of Effect
	1st	Awarded	1-8-19
	2nd	Awarded	2-10-20

An RFC service record for Ernest Frank Drewett (AIR 79/6)

The records cover those men who enlisted before First World War, including some Royal Engineers who had served in the Boer War, and men who continued to serve as late as 1924.

Arthur Edward Hobson enlisted in the RFC on 24 September 1912 as an Airman 2nd Class (RFC No. 403). He served in France with 2 Squadron between 1914 and 1916, being promoted to Corporal on 2 February 1915 and Sergeant on 2 May the same year. He rose to the rank of Sergeant Major and was mentioned in despatches for valuable services on 14 June 1918. He served at No. 6 Aeroplane Repair Section, 53 and 50 Training Depot Stations and in 120 Squadron. He was discharged on 23 September 1920.

Ernest Frank Drewett (RFC No. 404) joined the RFC on 2 October 1912 and continued to serve until 1 October 1924. His civilian occupation is listed as photographer so it is not perhaps surprising that by 1 April 1918 he had risen to the rank of Chief Mechanic (Photographer). He served in France between 16 August

1914 and 21 July 1916 and thereafter at various training schools and squadrons before being posted to India on 21 December 1921, where he served with 5 and 27 Squadrons and No. 5 Flying Training School. He was awarded the India General Service Medal for service in Waziristan, in addition to his three Great War medals and two Good Conduct badges. He was discharged from the RAF having completed twelve years' service.

For men who transferred into the RFC their previous service in other army regiments before joining the RFC is listed. James Richard McLean transferred to the RFC on 27 February 1918 but it is clear that he enlisted on 21 September 1914 with 6th Battalion, King's Liverpool Regiment. His medal entitlement shows that he served in France with them between 14 February and 27 July 1917. He is noted as having been admitted to Rouen General Hospital on 23 July 1917 and then to Portsmouth General Hospital five days later and then transferred to Merryflats War Hospital, Govan. He served as a driver in the RFC and was posted to Egypt in May 1918. He transferred to the RAF Reserve on 5 April 1919. Anyone interested in researching his whole career could follow his movements whilst with the 6th Btn King's Liverpools through their War Diaries which are in WO 95/1572 (February to December 1915) and WO 95/2926 (January 1916 to April 1919).

Records were also transferred for men who had previously been discharged, though here the information on a given individual may be small. The record for S Read (no first name given) shows that he enlisted on 29 April 1916 aged 34 years and 2 months, served in France between 16 October 1916 and 18 April 1917 and had spells in three military hospitals before being discharged under paragraph 392 XVI of the King's Regulations. There is no record of any next of kin, civilian occupation or unit(s) served in.

Frederick John Richardson died on active service with the RFC but his record too was passed to the Air Ministry. In his case his Short Service Attestation paper has been retained as a record of his service and it is neatly sandwiched between two records that have been transcribed in the correct numerical position (27617). His attestation shows that he enlisted on 29 April 1916 aged 28 years and 1 month, that he was married, lived at The Braes, Boyne Park, Tunbridge Wells, and was a motor mechanic. His wife was Alice Maud Caroline (nee Flatman) and he had a son, Frederick Richard, born 28 October 1908. He was promoted to Corporal on 1 April 1917 but on 31 December that year is recorded as 'Unposted believed drowned'. A later note records 'Drowned in loss of *HMT* [His Majesty's Troopship] *Osmaieh* 31st December 1917'.

Arthur George Banfield Rodgman (RFC No. 9940), of 33 Baker Street, Heavitree, Exeter, enlisted at Exeter on 13 October 1915. He was single, with his father being next of kin, and a tailor by trade. He was promoted Sergeant on 5 July 1917 and, at the same time, graded as a First Class Pilot at 63 Training School. On 20 August 1917 he was killed in an accident at 35 Training School. His Short Service Attestation is in AIR 79/126.

There is even a record for a man who enlisted in 1912 and died in 1914. Reginald George Cudmore (RFC No. 411) enlisted on 2 September 1912, declaring previous service in the Royal Engineers (Territorial Force). His Short Service Attestation papers show that he died on 15 May 1914. Further investigation shows that he was killed in the crash of a 2 Squadron aircraft at Northallerton.

For those men who served in the RAF Special Reserve there are service records in AIR 79/2807 and these occasionally contain details of service during Second World War.

Service Records for the South African Aviation Corps are in AIR 79/2806, the Attestation Papers for the South African Overseas Expeditionary Force.

What if my ancestor is not in AIR 79?

If you are unable to trace a record for a man you know served in the RFC during First World War, consider whether he might have continued to serve in the RAF after 1922, in which case his record will still be held by the RAF. Though most men were discharged within a year or so of the Armistice some did elect to stay on to become Regulars. Others may have been promoted to become officers (there was a surprising degree of this social mobility during the war, particularly in technical branches such as the RFC).

William Victor Strugnell (RFC No. 71) does not have a service record in AIR 79, but this is because he was promoted from the ranks (while serving as a Sergeant pilot with 3 Squadron) in 1915 and later served with the RAF in various capacities (mostly traceable through the RAF List) until he retired as a Group Captain in 1945. When he was initially promoted to be an officer he was commissioned into the Hampshire Regiment and there is an officer's file for him in WO 339/32887.

Though AIR 79 appears to be virtually complete and to run in numeric sequence there are occasional gaps which may be caused by a man transferring out of the RFC into another Army Corps or Regiment. A lot of fit men who served with balloon units were transferred to infantry battalions in 1917 for example. Surviving records for such men may be preserved among the army service records on microfilm at the National Archives in their WO 363 (the 'burnt documents') and WO 364 ('unburnt documents' – the reconstructed files) series. As the name suggests, very many (probably 60 per cent) of soldiers' records were destroyed by a bomb during the Second World War. The surviving records are organized alphabetically and if your ancestor survived and you know he was wounded, he may have been entitled to a pension, so start with WO 364. Even for unusual surnames it is likely you'll have to search half a dozen films to establish if the record survives. The record will usually have enlistment documents giving address, occupation and next of kin, as well as records of the man's movements and any wounds received. Some of the 'burnt documents' are, as you'd expect, in a dreadful state, having suffered smoke and water damage, though it is possible to find comprehensive sets of documents that weren't damaged at all.

TNA has begun to get WO 363 and 364 records digitally copied and put on-line.

RFC medical records

Casualty cards for the RFC and RAF (for men who died) are available at the RAF Museum, Hendon (see Appendix 9).

MH 106: officers

Though the hospital records released for other ranks in the MH 106 series are unlikely to be of any use unless you know when your man was hospitalized and which hospital he was sent to, the same is not quite true for officers of the RFC.

The series between MH 106/2202 and MH 106/2206 contains hospital records, organized alphabetically, of RFC officers wounded, giving not only name, rank, number and unit, but also a description of the medical problem. There are also, frequently, day-by-day descriptions of the process of the treatment, temperature charts and, in the case of wounds, a brief description of how these were received.

Please be aware that doctors' handwriting in the period was often as hard to read then as it is now!

Second Lieutenant E J Packe was wounded by a gunshot or shrapnel wound to his left buttock while flying at 2,000 feet above Beaumont Hamel on 1 July 1916 (the 'First Day of the Somme') and was evacuated back to Britain for treatment at the 2nd General Hospital at Bristol. An X-ray revealed a bullet lodged in his buttock above the hip joint, which was removed by an operation. His sutures were removed on 15 July and he was discharged to sick leave on 19 July.

Second Lieutenant A Pascoe of 13 Squadron RFC was wounded on 16 April 1917 while flying an aircraft over Arras as part of British operations in support of a huge French offensive launched further south on the same day. He was quickly evacuated to Wandsworth hospital in London where his wound was operated on thirty-six hours after he received it. There was some damage to the nerves in his arm and to his brachial artery but the wound was clean and by 18 May his wound had healed and movement of his fingers was improving.

The records are held in boxes and are loose, so great care is required when examining them.

MH 106/2202 Surnames A–C

MH 106/2203 Surnames D–H

MH 106/2204 Surnames I–O

MH 106/2205 Surnames P–S

MH 106/2206 Surnames T–Z

KILLED

ROYAL AIR FORCE. F.S. Form 558.

CASUALTY CARD. R. OF D. C H6394.

Surname HAINSWORTH 2/LIEUT Branch File No.

Christian Name GEORGE

Regiment R.A.F. Squadron

P. S.

Date of Report.	By whom reported.	Reference to List.	Date of Casualty.	NATURE OF CASUALTY.	Date published in Press.
6.12.18	ADMINISTRATOR COLLEGE GREENWICH	5.799	4.12.18	Letter 7/12 DIED. Pneumonia Queen Alexandra's Hosp Millbank N/K informed X	[illegible] 198
[illegible]	O.C. Queen Alexandra's Hosp Millbank	M.H. 3549/3.	23.11.18	Admitted influenza	
				KING'S LIST.	

P.T.O.

RAF casualty card for Lt George Hainsworth (1)

F.S. Form 559.

CASUALTY CARD.

Tests completed

111

Rank, Name and Unit Hainsworth 2Lt George
RAF att Training Estbt R.A.F. Greenwich from 35 T.D.S. Duxford.

Went overseas

At Time of Accident Employed as

Graduated as ______ on (date) ______

Date Report Received and Official Reference.	Date of Casualty.	Where occurred.	Type of Machine.	Nature and Cause of Accident.	Result of Accident.	Name of other Occupant of Machine.	Remarks.
F.S. 551. 6.12.18	4.12.18	Home		Died natural causes			Ad Queen Alexandra Mil Hosp, suffering from Pneumonia.

RAF casualty card for Lt George Hainsworth (2)

The official description says that the records cover the years 1916 to 1917 but some records do date from 1918. Second Lieutenant L H Pallister of 14 Squadron RFC was admitted to the Military Hospital at York on 13 February 1918 complaining of symptoms consistent with venereal disease and, after investigation, was discovered to have gonorrhoea. He was discharged to the Brighton Grove Military Hospital in Newcastle upon Tyne.

MH 106: other ranks

The MH 106 records are just a 5 per cent sample of the hospital and other medical units records that were made during the war. Most were destroyed in the 1920s but if your ancestor's service records shows that he was wounded or sick, and gives a medical unit, it is worth checking to see whether the documents survive. They are mainly admission papers, which will tell you a little about his (or on occasion her) condition, treatment, date of discharge and the unit to which they were sent. The earliest records date from April 1916 and extend to December 1919.

The surviving records cover field ambulance units (usually the first to receive a casualty after first aid treatment had been given by the Unit Medical Officer), casualty clearing stations (where casualties were sorted out for dispatch, if necessary, to hospital), general hospitals based in France or Belgium and hospitals in Britain where men were sent once they were judged capable of further travel. The Medical Services gave excellent treatment considering the state of knowledge at the time (there were no antibiotics of course) and, apart from the danger of infection, men stood a good chance of surviving injury if their treatment started early.

RECORDS THAT ARE KNOWN TO INCLUDE RFC, RNAS, RAF AND WRAF CASUALTIES

14 Field Ambulance, May–December 1918 (MH 106/59)

51 Field Ambulance, August 1918 (MH 106/121)

66 Field Ambulance, June–October 1918 (MH 106/153 and 154)

139 Field Ambulance May–September 1918 (MH 106/205)

3 Casualty Clearing Station, June 1918–January 1919 (MH 106/392 & 393)

11 Casualty Clearing Station, May 1918–March 1919 (MH 106/513)

31 Casualty Clearing Station, May 1918–September 1919 (MH 106/645, 647, 652 and 665)

34 Casualty Clearing Station, June 1918–January 1919 (MH 106/742, 793 and 794)

39 Casualty Clearing Station, May 1918–March 1919 (MH 106/810 and 811)

2 General Hospital, May 1918 – March 1919 (MH 106/1029, 1030, 1034, 1035 and 1037)

18 General Hospital, July 1917–January 1919 (MH/106/1139, 11440, 1141, 1143, 1145, 1146, 1147, 1148, 1149, 1153, 1156, 1157, 1164, 1180, 1187)

19 General Hospital, April 1916–November 1918 (MH 106/1235, 1238–58, 1259, 1260, 1261, 1264, 1265, 1266, 1269, 1270, 1281, 1284, 1288, 1290, 1291, 1292)

28 General Hospital, August 1916–April 1919 (MH 106/1316, 1318, 1322, 1333, 1335, 1337, 1339, 1340, 1365, 1368)

85 General Hospital, June 1918–April 1919 (MH 106/1368)

4 Stationary Hospital, October 1916–September 1919 (MH 106/1460, 1483, 1491, 1494, 1497)

Middlesex County War Hospital (Napsbury), March 1918–July 1919 (MH 106/1528–30)

Queen Alexander's Military Hospital, Millbank June 1917–May 1919 (MH 106/1636, 1637, 1638, 1639, 1690, 1691, 1692)

Catterick Military Hospital, January 1918–December 1919 (MH 106/1825 and 1899)

Craiglockhart Hospital, June 1918–January 1919 (MH 106/1900, 1901 and 2037)

31 Ambulance Train, June 1918–April 1919 (MH 106/2038 and 2040)

It must be stressed that other hospital and medical units *may* contain records of your ancestor and you should try searching for the relevant unit and period using the search facility of TNA website if you know where he was treated.

Royal Flying Corps unit records: the structure of the RFC and its influence on the records

The smallest RFC (and later RAF) unit is the Flight, consisting usually of three of four aircraft. These are usually grouped into Squadrons of three or four Flights, but some Flights do operate independently. For command purposes the Squadrons are grouped into Wings (usually of three or four Squadrons) and the Wings into Brigades (renamed Groups when the RAF was created). After the creation of the Royal Air Force the Groups were organized into Commands, first based on geographical location and later on the kind of work that the aircraft were expected to do, i.e. Fighter Command and Bomber Command. Have a look at some of the Orders of Battle in Appendix 2 to see how the structure evolved and how it worked.

Unless a Flight is operating independently its records will be in the Squadron records. Squadron records will include reports from individual pilots and observers, combat reports and reports on individuals and on aircraft. Some of these reports were copied to Wing Headquarters who would then forward some of them on to Brigade/Group. This does mean that some reports missing from Squadron records may be found by looking higher up the chain of command. There are some sample Orders of Battle for the RFC in the appendixes which may show which wing and group a squadron you're interested in belonged to.

War Diaries

Following the problems that it had faced in the Boer War the army conducted a thorough review of what had gone wrong and how it thought things should be corrected. One of the first problems it addressed was that unit record keeping varied considerably, and it was difficult to analyse what had been done in retrospect. As a result, War Diaries, written (at least theoretically) every day during wartime, were introduced. They were to record unit movements, orders and events of note. Standards of record keeping varied and, at times of pressure, such as the retreat from Mons in 1914, it is clear that the duty was neglected due to the simple pressure of keeping moving. Being part of the Army every RFC Squadron was obliged to keep a War Diary and they are an invaluable source of information for the early part of the war. They were replaced, at Squadron level, by Squadron Record Books at the end of 1915, but Wings and Brigades continued to keep them until well into 1918.

To France with the British Expeditionary Force (Squadron War Diaries)

On 13 August 1914, just over a week after war broke out, 4 Squadron's aircraft were at Eastchurch, the aircraftsmen and ground equipment having embarked for France overnight. The War Diary records:

> Prepared to start at 8.30. Got machines out at 6 a.m, ran engines, breakfasted, delayed for fog to clear, sent Playfair for rifle grenades, could not find where Hardelot was. Ordered Shepherd to start first go to Gris Nez and south along the coast from there, then find ground, land and put down strip, others to follow but not to land until strip was down.

Shepherd, with Bonham-Carter as his observer, crossed the Channel between 5,000 and 6,000 feet and landed at Berck at 11.30 and having been unable to find Hardelot, the village that was supposed to be their rendezvous with the rest of the squadron, they flew on to Amiens and arrived at 3.15. The other pilots had mixed luck: Cogan, having crossed the Channel successfully, had an engine failure and landed at Equihen and broke his landing skids; Roche too had problems and landed at Equihen but made it to Amiens just after 1.15; Morgan saw Cogan on the ground at Equihen and landed to help, breaking his undercarriage. The other officers, some having to land for spares or fuel along the way, made their way to Amiens successfully and obtained comfortable billets in the Hotel de l'Ecu de France. The men slept under the machines in a field.

The other officers came in one by one and on 16 August the squadron made ready to follow the BEF into Belgium. Early morning mist being slow to clear, at noon 'machines started for Maubeuge at 1 minute intervals'. By the evening of 17 August the aircraft were at Maubeuge aerodrome with 2, 3 and part of 5 Squadron. Within a few days they were to be seriously at war.

On 20 August (according to 3 Squadron's diary) they heard about the fall of Liege to the Germans. 'Five German Army Corps E of R Meuse & facing French forces disappear. Supposed intention a wide turning movement on Brussels. German advance guard in Brussels.'

On 21 August, though reports were coming in of large-scale German advances, there was little sign of these to the RFC patrols. Captain Herbert of 3 Squadron flew a long patrol with Major Moss as observer and reported 'All country clear of troops with exception of a few cavalry patrols near Soinglies'. On 22 August Lieutenant Wadham, with Captain Charlton observing, saw

> infantry in small columns discernable at Hal and neighbouring towns. Landed for enquiries at Hoerberke (2 miles SE Grammont) and learnt about 5,000 Germans in Grammont; cavalry and cyclists in Lessines. Cavalry expected in Ath this evening, all from Enghien direction. Re-ascending was fired on by about an infantry brigade with guns halted at Bassily.

On 22 August the squadrons carried out twelve scouting patrols and began reporting huge numbers of Germans moving in on the BEF as it settled down at Mons. An aircraft of 2 Squadron came under groundfire at Maffle, south-east of Ath, and Sergeant Major Jillings was wounded in the leg (see section on casualty cards). An aircraft of 5 Squadron took off on patrol and never returned, Second Lieutenant Waterfall and Lieutenant C G Bailey being reported as missing in action. A German aircraft also passed over Maubeuge airfield and was chased off by three aircraft, one mounting a Lewis gun (a light machine gun) that slowed it down so much it was ordered to be removed.

On 23 August the whole army began the retreat from Mons, falling back daily in front of an enemy that outnumbered them several times over. On 24 August Captain Shepherd and Lieutenant Bonham-Carter observed German armies moving to outflank the whole BEF and their report convinced Field Marshal French to continue the retreat.

There is a nominal roll of all the officers and men who went to France with the RFC in August 1914 in AIR 1/765/204/4/237. This shows which squadron or other unit the men were attached to (there was an aircraft park and some officers on staff duties, for example) and, in the case of the officers, it gives their original regiment and their rank within the RFC as well as their regimental rank (which might be different).

Balloons

As well as aeroplanes and airships both the RFC and RNAS were responsible for the use of balloons, both 'free', i.e. moving balloons, which were generally used for training, and 'fixed', i.e. secured to the ground or to a ship, which were used for both observation and as a deterrent to aircraft.

The main type of fixed balloon was the 'kite' balloon which had a tail and would thus turn into the wind and sit fairly stable, giving a fixed platform for the observers. Based a mile or two behind the front line a balloon could rise to up to 5,000 feet and provide the observers in the basket with the opportunity to watch developments behind the enemy lines as they occurred so they could telephone news down to artillery batteries. Balloons became a prime target for enemy artillery and fighter aircraft and, being filled with inflammable hydrogen gas, were very vulnerable. The crew were equipped with parachutes so that, in the event of attack, with the ground crew winching the balloon down frantically, the crew could leap over the side and descend independently.

Free balloons were used mainly for training officers for the fixed balloons so they would know what to do if the cable was cut, but occasional attempts were made to use balloons to send propaganda and agents across the front. A particularly daring attempt involved sending an agent into Luxembourg in 1918, beautifully researched and written up by Janet Morgan in *Secrets of the Rue St Roch* (Allen Lane, 2004).

By 1916 the balloon units were grouped into wings, each wing covering the front of a whole army, so that 1st Balloon Wing covered 1st Army, 2nd Balloon Wing covered 2nd Army, etc. Each wing comprised several balloon companies, each of which comprised, at first, two balloon sections, though this number rose as the war progressed. Balloon units served on most fronts where there was static fighting, including in Salonica, Egypt, Mesopotamia and North Russia.

Records for the various balloon units vary. Some wings, companies and sections seem to have produced short histories before they were disbanded in 1919 and there is a good selection of these between 2 Balloon Wing History (AIR 1/163/15/126/1) and a history of 39 Balloon Section (AIR 1/163/15/135/1). The papers relating to 5 Balloon Company (AIR 1/1950/202/257/1–AIR 1/1950/204/257/9) contain particulars of officers, flying times of officers and other ranks, reports on attacks by hostile aircraft, recommendations for awards, field returns and the flying times log book.

The RNAS used kite balloons towed by fast-moving ships such as destroyers for spotting for the fleet and for anti-submarine surveillance. As well as in the AIR 1 files there is material on some of these units in ADM 1/8463/177, ADM 137/327 and ADM 131/64.

The RFC (and later RAF) were also responsible for the barrage balloons protecting London. First seriously suggested as a means of defending the capital in 1916, work on a series of lines of fixed balloons, connected by cables and trailing 1,000-foot-long steel wires, was then begun. Ten lines of defence were completed by the end of the war. The balloons could reach a height of 10,000 feet and their purpose was effectively to close areas of the sky to enemy aircraft and force them into the path of fixed patrols of British aircraft, or close to batteries of anti-aircraft guns. The balloons were not kept airborne the whole time but were raised in anticipation of air raids. As a result, small detachments of RFC men were scattered around the outskirts of London and out along the Thames estuary, manning the winches to raise and lower the barrage.

There are various files on the working of the balloon barrage (or 'Apron') throughout AIR 1. AIR 1/2051/204/379/3 relates to the initial proposals; AIR 1/609/16/15/271 details some of the sites that the balloons worked from. AIR 1/610/16/15/276 includes Field Marshal Lord French's report on the defence of London during 1918. AVIA 20/247 is a general report on balloon defences, including the London Barrage.

Records for both officers and men are held with the main records for RFC/RNAS/RAF officers and men. There are some nominal rolls for individual units in AIR 1.

Captain (later Brigadier) P J Slater was in command of No. 28 Balloon Section RFC/RFC and donated his photograph album to the Museum of Army Flying (see Appendix 9), which also holds an extensive library on military ballooning.

An excellent description of the life of a balloon officer, his training and observational work, as well as helping with the London Barrage, can be found in Goderic Hodges's *Memoirs of an Old Balloonatic* (William Kimber, 1972).

Squadron Record Books (First World War)

The earliest Squadron Operations Record Books (ORBs) date back to at least 1913 (5 Squadron's ORB from July 1913 onwards is in AIR 27/63) and the majority of them for the First World War are in AIR 1. The practice of keeping Squadron War Diaries seems to have ceased (with the exception of the South African Squadron) around about the end of 1915, though Wing and Brigade Diaries continued. The original instructions, as issued by RFC Headquarters, on how to keep record books is in AIR 1/816/204/4/1287.

Record books are designed to be a record of the 'working' side of the squadron, recording the flights made and the people who made them, and were kept on a daily basis. A typical book will record the type and number of the aircraft involved, the pilot and observer, the purpose of the flight, start and return times, numbers of hours in the air and a general section called 'Remarks'. Here the aircrew would note the success (or otherwise) of their mission, any enemy activity seen, any bombs dropped, how many photographs taken, etc. On 3 October 1918 34 Squadron, then fighting alongside the Italians in Northern Italy, sent Captain H A Pearson MC (pilot) and Lieutenant S Jackson MC (observer) on a photographic reconnaissance of the front line in RE8 E/263. In a successful flight lasting 2 hrs 30 mins they took 19 exposures of the Corps front, of which 18 were successful. No E.A. (enemy aircraft) were seen and hostile anti-aircraft fire was normal. Visibility was good and the flight was made between 7,500 and 10,000 feet.

You will frequently find other kinds of report inserted in the record book, copies of reports made by the pilot or observer about a particular mission flown. On 29 October 1918 Lieutenant Etheridge (pilot) and Lieutenant Lovell (observer) flew a cavalry contact patrol in a 34 Squadron RE8 to observe, and keep in touch with, troops advancing behind the Austrian–German army, which was then in full retreat.

They were debriefed by an Intelligence Officer on their return and gave a full report of what they had seen and done. Amongst other things they had observed

Whoops! The fate of a 'Harry Tate'. Lt Patterson's RE8 after a collision with some mulberry trees, Italy, November 1918.

And again from another angle! (With thanks to the Patterson family).

An RE8 of 34 Squadron in flight, Italy, November 1918. (With thanks to the Patterson family)

'Cavalry Patrols advancing on general line RAI – C. GRASSALOTTO – C. BARTOLLETTI – CALE.' At 12.30 hours they saw British infantry over the River Monticano above Vazolla, Italian troops advancing near Sarano and anti-aircraft fire near Santa Fior di Sopra. They dropped messages at 23rd Division HQ and at a Corps Dropping Station and saw only one hostile aircraft.

The following day Lieutenants Grosset (pilot) and Rankin (observer) flew a similar contact patrol and were able to report British infantry and cavalry advancing, Italian units moving forward, and big fires behind enemy lines. Clearly the enemy was in full retreat.

Even after the Armistice squadrons continued to record flights made, so that one can see officers training, moving aircraft and testing equipment on a day-to-day basis.

Unit histories

AIR 1 contains numerous unit histories, not only of individual squadrons but also brigades, balloon companies, individual flights (where operating independently), groups, depots and bases. Some squadron histories were written at the end of the war, but others were compiled in the 1930s when the RAF began its expansion in anticipation of the Second World War. If you're lucky there will be two histories, as there are for 34 Squadron! AIR 1/173/15/163/1 was written in 1919 and covers the years 1916–19. AIR 1/691/21/20/34, though it nominally covers the squadron's history from 1916 to 1937, is much more detailed on the early period.

> The Squadron joined about 7th January 1916, with a nucleus flight of 11 men from No. 19 Squadron at Castle Bromwich. A few days later Captain J A Chamier, 33rd Punjabis, I.A., attached RFC, arrived to take command.

The history gives brief details of training and mobilization for war.

> On July 10th 1916, the Squadron moved to France – transport proceeding via Avonmouth, Rouen, to Hesdin without a hitch. The aeroplanes flew over and arrived complete in three days without damage.

The squadron had already put up several records:

> (1) It was the first squadron to land all eighteen aeroplanes and all transport at its destination without a crash.
>
> (2) It was the first squadron on which the Flying Officers averaged fifty hours in the air.
>
> (3) It was the first squadron of which every pilot had looped before proceeding overseas.

The squadron was assigned to work as Corps Squadron with 3rd Corps, 4th Army during the Battle of the Somme where it 'rapidly gained for itself a high reputation as a Corps Squadron, particularly in artillery cooperation; it helped to develop this and Contact Patrol into a system, and saw the first use of tanks from the air'. During 1917 the squadron fought ninety-eight combats during 1917 but suffered only two fatal casualties, five officers wounded and two taken prisoner. 'The lightness of the casualties were due to the handiness of the B E 2e and the skill in "stunting" of the pilots due to their "hours in the air" before coming overseas.'

The AIR 1/173/15/163/1 history gives details of all the squadron movements from formation in October 1916 at Castle Bromwich to their time on the Somme in 1916, their move to Italy in November 1917 and their bases there. It explains which Brigades they served under and who their Commanding Officers were. They destroyed six enemy aircraft and drove six down out of control, dropped approximately 92 tons of bombs and took 18,132 photographs.

The history lists thirty separate honours and awards to officers and twelve to other ranks, as well as all officers and men killed, died of wounds, died of injuries, taken prisoner, wounded by the enemy and otherwise injured.

First World War combat reports

Combat reports were written by the officers involved as part of their debriefing after they had landed and are designed to capture an immediate picture of the fighting they had been involved with for later analysis by the Squadron and Wing Intelligence Officers. A typical report from AIR 1/1219/204/5/2634, part of a large collection of combat reports, reads:

> Was returning from taking photos of an AA battery and a balloon winch near Mericourt, when we saw three Fokkers with a type 'B' in the rear approaching from the ARRAS direction. Just as two of them were circling to drive at us from the rear I turned sharply towards them. The observer got off ½ a drum at the first. The rest of the drum he fired at second Fokker. Some of the last shots hit the front of the machine.
>
> I then turned towards our lines to allow the observer to load. Just then

2 THE BAT.

LIST OF OFFICERS

AT

R.A.F. STATION, GREAT YARMOUTH,

28th FEBRUARY, 1919.

Lieut.-Col. E. T. R. Chambers
(Commanding Officer)

Capt. C. H. James
(Adjutant)

Major E. Cadbury *(O.C. 212 Squadron)*	Major R. Leckie *(O.C. 228 Squadron)*	Major A. B. Gaskell *(O.C. 229 Squadron)*
Capt. A. E. Siddons Wilson *(O.C. Wing Repair Section)*	Capt. V. H. Ridewood *(Wing Armament)*	Capt. J. C. H. Allan *(M.O.)*
Capt. H. Wade *(M.O.)*	,, E. A. Bolton	,, C. B. Sproat
,, L. C. Keeble	,, C. D. Kirkpatrick	,, C. McCann
,, C. Wincott	,, F. C. Vincent	,, P K. Fowler
,, B. C. H. Cross	,, H. W. Buckley	,, A. H. Sole
,, Inge	,, S. J. Featherston	,, A N. FitzRandolph
	,, J. Hodson	,, A. C. Teesdale
Lieut. O. W. Pellatt	Lieut. H. G. Owen	Lieut. R. E. Keys
,, H. F. Potter	,, W. H. Cumming	,, M. P. Pearson
,, R. Ellingham	,, W. H. Bicknall	,, H. A. Brosse
,, H. W. Basedon	,, H. Bricker	,, A. Wroote
,, F. Burton	,, C F. S. Gamble	,, C. F. Wagstaff
,, S. H. Bazeley	,, R. Howard	,, G. R. Halliday
,, H. Whates	,, A. D. Pole	,, L. Plater
,, G. H. Bloom	,, G. R. Burge	,, K. B. Preston
,, J. E. Monypenny	,, V. S. Green	,, G. S. Musson
,, E. E. Crook	,, F. R. Walpole	,, W. B. Cass
,, A. P. Bell	,, R. Grey	,, L. W. Wilson
,, W. H. Comstock	,, L. Butchart	,, B. Barwell
,, A. C. Hands	,, H. K. Johnson	,, W. J. D. Langlois
,, A. Macfarlane	,, H. P. Guard	
2nd Lieut. W. H. Heywood	2nd Lieut. O. D. Bell	2nd Lieut. W. R. Waterman
,, J. Harding	,, M. C. Horkins	,, F. H. Stringer
,, C. Martin	,, G. W. C. Ravenhill	,, H. G. Harris
,, R. K. Stinson	,, F. H. Devlin	,, C. D. Frogbrook
,, H. R. Gomershall	,, S. Toby	,, M. T. Stanley
,, A. J. Colvin	,, W. Wilson	,, C. H. Abbey
,, E. Grahame	,, H. May	,, R. N. Hesketh
,, H. E. Power	,, J. F. Young	,, R. H. Boyden
,, C. H. Smith	,, G. A. Meek	,, G. Fisher
,, L. L. Bridgeman	,, T. W. Deary	,, W. Littlemore
,, H. Walter	,, F. H. Whitmore	,, R. E. H. Allen
,, J. B. Harper	,, J. Roberts	,, C. C. Woods
,, H. C. Armitt	,, M. J. Hennessey	,, S. H. Thornton
,, E. B. Hyndman	,, W. C. Stevenson	,, F. Willis
,, T. W. Oliver	,, G. R. Grain	,, R. H. Galloway
,, J. Macdonald	,, D. H. W. Crossley	,, F. Gandy
,, J. F. F. Liddy	,, F. Eppinger	,, F. Cartwright
,, C. W. B. Llewellyn	,, L. W. Kidd	,, G. A. Evans
,, R. W. Willis	,, W. C. Ingram	,, J. Greenwood
,, E. Jolley	,, C. P. King	,, A. O. Osborne
,, J. F. Dodd	,, G. E. Lewtass	,, T. Ginn

A list of RAF officers at RAF Great Yarmouth, January 1919, taken from the station magazine 'The Bat'. (Author's collection)

> the third Fokker firing at us from the rear put the engine out of action. I put her nose down towards our lines. The other two Fokkers had turned and were again firing at us from the rear. I told the observer to crouch down. Their shooting was most accurate bullets kept tearing past me on both sides. Some of the shots took away my elevator controls and left me without any fore and aft control. Over the lines the firing stopped. I looked into the front seat and saw the observer lying over to the right of the nacelle with a ragged bullet hole in his skull. In crouching down he had evidently put his head on the right side of the nacelle and it was one of the shots that went past my right arm that killed him. I let the machine make her own gliding angle and affected a fair landing opposite the CHATEAU at La Hare.
>
> (Signed) J C Callaghan
> G I Carmichael Major RFC

One unfortunate point to note about the combat reports from both the First and Second World War is that a large number were stolen from the Public Record Office in the 1980s and, though some were recovered when the individual was caught, many were not. If approached by anyone selling or offering original combat reports (or other RFC or RAF records) do take extreme care and if not absolutely satisfied as to their bona fides report the matter to TNA. The good thing is that many reports were duplicated as they were circulated to Wing and Brigade level so a search of records at that level may produce copies that you couldn't find at Squadron level.

Putting the records together: the RFC on the first day of the Somme

At 7.30 a.m. on 1 July 1916, as an artillery bombardment that had already lasted a week reached its crescendo, thousands of British and French troops advanced on the German lines and began the Battle of the Somme. It was the biggest battle so far fought by the British armies in France and the biggest so far for the Royal Flying Corps. The principal attack was to be made by General Rawlinson's 4th Army supported by further attacks in the north of the Somme sector by General Allenby's 3rd Army.

The Battle of the Somme could be said to have begun, for the RFC at least, considerably earlier than July, because the British Army had only taken over the sector they were to attack at the start of the year, and much photographic reconnaissance had to be done in anticipation of the attack. Bombing of railway junctions behind German lines was commenced, and every attempt was made to draw the German Air Service into battle so as to destroy them and achieve air superiority. Once the ground attack had commenced the six squadrons of 4 Brigade were quickly reinforced by four squadrons of 9th (Headquarters) Wing, giving a total of 167 aircraft, 76 of which were new fighters.

Instruction 29 by Brigadier General E B Ashmore, CMG, MVO, Commanding 4 Brigade, Royal Flying Corps, dated Friday 30 June 1916, reads:

> On tomorrow, July 1st, the 14th Wing will continue to patrol the line from daylight to dusk as directed . . . except between the hours of 6.45 am and 10.15 am when the front of the Fourth Army will be patrolled by 4 FE's and 6 scouts from the 14th Wing continuously.

> O.C. 14th Wing will arrange for offensive patrols in addition to the protective patrols after 10.15 am.

The weather on 1 July was very fine and bright all day, though with some early morning mist which obscured the view of the assault for the kite balloons of No. 1 Kite Balloon Squadron (specially reinforced by additional balloon sections for the battle). This was important because the balloons, with their telephone lines straight to the ground, were the army's best source of immediate information. The mist also obscured the immediate front from the reconnaissance aircraft, at least until it burnt off later in the morning.

The Squadron Record Book for 9 Squadron (AIR 1/1233/204/6/13) shows that the first reconnaissance flight was up at 4.10 a.m., with Second Lieutenants D Hall and A Wynn carrying out artillery observation, but a very thick mist made the accuracy of bombardment impossible to observe. Throughout the day the squadron's BE2c aircraft continued to fly up and down the line watching for troop movements and directing the British artillery onto hostile positions. At 07.30 Captain V Robeson and Second Lieutenant G Stapylton saw 'Infantry seen to advance except at line through A7 and A8. The flashes of bayonets seen in Dublin Trench. Infantry seen about 40 yards from Pommiers Trench.'

The British, often accused of learning little from their failures, had in fact learnt a great deal from earlier battles such as Loos, where troops had advanced out of line of sight of their senior officers and become cut off from communication with Headquarters. Soldiers wore small, bright tin plates on their backpacks so that they could be seen from the air and carried panels that could be laid out on the ground to communicate with the spotting aircraft. They also carried flares and semaphore devices with which they could pass basic messages to the aeroplanes. Artillery fire was arranged in simple zones so that, by giving a combination of letters and numbers to the artillery batteries, the aircraft could arrange for fire to be switched rapidly onto new targets.

Aircraft from 9 Squadron watched and reported the advance of the three battalions of 54th infantry brigade into Pommiers Redoubt. Second Lieutenants A Macdonald and H Hill

> Dived to 900ft over Pommiers Redoubt. A white panel was distinctly seen about 25 yards N of Redoubt, bayonets were also seen in redoubt and about 50 yards to the North.

At about the same time (10.15 a.m.) Second Lieutenants I Macdonell and H Williamson saw the British infantry reach the outskirts of Montauban and saw

> About half a company German Infantry in Quarry at S 22 c 0050, formed up in small squads. Dropped two bombs on them from 1800 feet and fired one drum from machine gun. Ranged 15 Siege Battery on them successfully.

Captain J Whittaker and Second Lieutenant T Scaife reported: 'Our infantry (mirrors on backs) seen in Train Alley trench. Men seen leaving Glatz redoubt (with mirrors) along whole trench. Our men in splendid formation seen going into Montauban.' Having established that in their sector the troops had achieved their objectives the squadron flew deeper behind German lines, reporting movement of troops and transport and flying until 9.45 p.m.

Most other IV Brigade Squadrons record books from the period do not seem to have survived, but there are summaries of their reports in the 4th Brigade War Diary (AIR 14/2248/209/43/12). The summary of daily events reveals nine aerial combats during the day, with one German machine driven down, several bombs dropped on enemy batteries by 4 Squadron, many valuable Contact patrols made, several batteries knocked out, a fair number of photographs taken and two FE2b aircraft missing. The Diary records events as they were reported to Brigade Headquarters so there can be quite a difference in timing between what was happening on the ground and what the senior officers knew about it (one of the major problems on the day, which resulted in follow-up attacks being launched before it was clear that the initial attacks had failed). 9 Squadron first spotted the capture of Pommiers Redoubt at about 7.30 a.m. but it is not reported in the War Diary until 9.20 a.m. 4 Squadron reports about the fighting around La Boiselle, in which the Northumberland Fusiliers were cut to pieces by machine gun fire, did not reach Brigade Headquarters until 8.25 a.m. A report from No. 1 Kite Balloon Squadron, which had balloons up all along the front, was inconclusive. At 8.45 a.m. they reported: 'Low cloud over trenches: observation very difficult. No flashes seen. One balloon saw our infantry leave trenches. German barrage seen in places; from its position we seem to have moved forward.' At 9.45 4 Squadron reported '13th Royal Irish Rifles held up by machine gun fire.' At 10.10 a.m. Lewis of 3 Squadron reported British infantry 'close in front of Pozieres' but the diarist records 'I told this to Army and Army Commander. They are a little doubtful whether it can be true.' In fact the attack by 2nd Middlesex and 2nd Devonshires towards Pozieres had barely reached the German front line, let alone the spot reported by 3 Squadron, and were soon ejected from the captured trenches by a German counter attack. At 11 a.m. a 15 Squadron contact patrol reported 'our men in first and second trenches all along the sector'. In one of the few air battles of the day Cowan of 24 Squadron 'fought 2 at once, disabling observer of first, and 2nd fell into a cloud out of control'. A contact patrol of 3 Squadron reported at 11.36:

> 7.45 Infantry crossing No man's land, X.27.c and at F.3.a. 7.55 Flares seen in groups North of Fricourt, and a few scattered flares South of Fricourt. No flares or ground sheets seen in III Corps. It appears that our infantry were on the line of trenches in front of Pozieres. They had not reached Contalmaison Wood nor Contalmaison. Information dropped by message bags to XV Corps.

Throughout the day the aircraft of 4th Brigade spotted German artillery batteries and directed counter-battery fire onto them from the British guns, reported German troop and transport movements and the positions held by British troops. Unfortunately there were frequently delays in getting the information back. 9 Squadron reported that between 12.45 and 2.30

> Battalion ground sheets at (1) junction of Pommiere and Lane trench (2) Pommiere Lane near Red Redoubt (3) junction of Casement trench and Glatz Alley (4) Austria junction. Infantry commenced to advance SE of Fricourt and appeared to reach German second line.

This report is logged into the Brigade War Diary at 7 p.m. The last report, timed at 8.43 p.m., was entered into the diary at 9.40 p.m.

Shortly afterwards Brigadier General Ashmore wrote a brief report on contact patrol work that the aircraft of his 4th Brigade had carried out between 1 and 4 July. He noted that the

> mirrors on men's backs were very clearly seen in the XIII Corps attack. On this day the attacks were very clearly followed from the air, but communications from the infantry were not frequent, probably because the telephone system on the ground was insufficient.

In fact communications on the ground had generally been bad, with many telephone wires cut by the German barrage. This may explain Ashmore's later remarks that

> After the first day, flares, ground sheets and panels have been of the greatest use. For example, after the assault on Bernafay Wood in the late evening of July 3rd Battalion HQ showed their ground sheet, sent their code call, and were reported to the Corps Artillery HQ by wireless, within a few seconds. Again, the position of our troops at Ovillers and in La Boisselle was constantly defined during the 3rd by flares, and by flares only.

In spite of training and practice, some things only really work when they have been tested in the heat of battle and shown to stand up to conditions. With the confidence that came from knowing that the RFC could spot and report their markers and messages the troops were more inclined to use them, and cooperation between the infantry and RFC improved markedly as the battle progressed.

One aspect of the day that was considered a success was the tactical bombing of German transport behind the lines. This was in the hands of 9th Wing RFC under the command of Lt Colonel Hugh Dowding. In a later report he wrote

> The raid (on Lille) started soon after 4 am and was successful. The raid was carried out by six R.E.'s escorted by tow Martinsydes and two Morane biplanes. A further escort consisting of five F.E.'s from another Brigade was engaged by a Squadron of Fokkers before it had effected a junction with the machines of 9th Wing. I was informed that they had fought a successful action with these Fokkers bringing two of them to the ground in the German lines. The effect of this was that the raid was almost unmolested by aircraft, though one Fokker succeeded in diving ahead of the escort and attacking the leading bombing machine. Material damage was done to the machine but the pilot was not hit. The morning was hazy, and it was difficult to observe the effect of the bombs, but it is believed that three 336 lb bombs exploded in the station buildings and sidings. All the machines returned safely.
>
> The first bomb raid on Bapaume was carried out by six Martinsydes escorted by two Martinsydes and 2 Morane Scouts. The machines left the ground shortly before 8 a.m., and material damage was done to Bapaume where a fire was started that was not extinguished for several hours. The expedition was considerably molested by hostile aircraft and the coolness of Lieut. J C Turner (No. 27 Squadron) is worthy of mention. This officer's engine was running badly and one of his bombs failed to release itself at the first attempt. Although continuously attacked by two German machines, this officer made two more circuits over his objective until he was finally able to release the bomb by using both hands. By this time his machine had descended to five thousand feet and he re-crossed the line at two thousand five hundred feet.

A further attack was made on Bapaume later in the day, and reconnaissance was also flown of the railway at Busigny. The escorting machines were continuously attacked by German aircraft, two of which were forced down by Second Lieutenants S Dalrymple and H A Taylor of 27 Squadron. Offensive patrols also forced down a German LGV Biplane over Bapaume.

Dowding noted

> The general impression of the day's work has been that the Germans started with the intention of attacking any of our machines which might cross their lines. As the day wore on they were gradually driven from the sky, and after 1.00 pm scarcely a German machine was seen in the air, and those which were manifested no aggressive tendencies.

Whether he appreciated it at the time, Dowding was learning the importance of air superiority that he was to carry forward successfully in stalling German plans for the planned invasion of Britain in 1940, when Dowding led Fighter Command during the Battle of Britain.

At the end of 1 July 1916 the whole British attack had stalled in the northern sector, but advances had been made in the south, in the area covered by 4th Brigade RFC. 30th Division had captured all its objectives and 18th, 7th and 21st Divisions had made significant gains. The French further south had also advanced successfully. General Rawlinson began to shift the focus of the attack south and over the next few days more advances were made. On 14 July 4th Army captured 6,000 yards of the German defences between Longueval and Bazentin-le-Petit in an imaginative dawn attack accompanied by a creeping barrage. 4th Brigade aircraft covered these advances, developing the techniques of artillery and infantry cooperation.

During the course of the first two weeks of the fighting the RFC gained almost complete aerial domination over the battlefield, denying the Germans reconnaissance and artillery direction. German reinforcements were rushed up, including new aircraft, so that by the beginning of August numbers were almost equal between the two air services. By the end of September the Germans succeeded in fighting the RFC to a standstill, killing sixty-two aircrew, wounding thirty-six and forcing down another thirty-six to be taken prisoner. To assist the RFC the Admiralty were even prevailed upon to loan RNAS squadrons, the first of which, 8 Squadron RNAS, arrived on the Somme in November. Several more squadrons were transferred during the next few months, their generally superior aircraft (Sopwith Pups, Triplanes, 1½ Strutters and Camels) taking on the German fighters on an equal basis.

As the battle developed, heavy fighting continued in almost unimaginable conditions around Delville Wood and High Wood, with the British trying to develop a foothold in the German lines for another major offensive. South African, Canadian, Australian and New Zealand troops were thrown into the fight. The newly invented tanks were used in the capture of Flers on 15 September as part of a major new offensive which scored great success but still didn't break the German lines. Heavy fighting continued, in deteriorating weather conditions, until 19 November.

Records of the RFC on the Somme

Records of the fighting over the Somme on 1 July 1916 are scattered, with many apparently missing. Some of the more important are:

4TH BRIGADE

AIR 1/2248/209/43/12 War Diary, 1916 June–Sept.

AIR 1/759/204/4/137 Work summary of 1 Wing, 9 Wing and 4 Brigade RFC 1916 June–July

AIR 1/765/204/4/243 Summary of Brigades, Wing and Squadron work, 1916 July

3 Squadron

AIR 1/687/21/20/3 History of 3 Squadron, RAF, 1912–1921

AIR 1/1216/204/5/2634 Combat reports: 3 Squadron 1915 May–1918 Oct.

4 Squadron

AIR 1/1217/204/5/2634 Combat reports: 4 Squadron, 1915 May–1918 Apr.

9 Squadron

AIR 1/688/21/20/9 History of 9 Squadron, RAF 1915–1937

AIR 1/1233/204/6/13 Record book, 1916 July

AIR 1/1234/204/6/19 Record book. Volume 2. 1916 July–Aug.

AIR 1/1240/204/6/47 Record book, 1916 June– July

AIR 1/1243/204/6/59 Daily routine orders, 1916 May–Aug.

AIR 1/1246/204/6/72 Short history, 1915 Apr.–1918 Nov.

15 Squadron

AIR 1/689/21/20/15 History of 15 Squadron, RAF, 1915–1919

AIR 1/1219/204/5/2634 Combat reports: 15 Squadron, 1916 Jan.–1918 Nov.

AIR 1/1359/204/21/9 Combats in the air, 1916 Jan.–1918 Nov.

AIR 1/1359/204/21/10 Reports of officers: various subjects, 1916 Jan.–1919 Feb.

13 Balloon Section

AIR 1/163/15/130/1 History of 13 Balloon Section, RAF 1916–1918

22 Squadron

AIR 1/1220/204/5/2634 Combat reports: 22 Squadron, 1916 May–1918 Nov.

24 Squadron

AIR 1/690/21/20/24 History of 24 Squadron, RAF, 1915–1936

AIR 1/1221/204/5/2634 Combat reports: 24 Squadron, 1916 Apr.–1918 Oct.

AIR 1/844/204/5/372 Reports on aeroplane and personnel casualties, July 1916

AIR 1/895/204/5/712 Photography: Somme Battle ground, 1916 July–Dec.

AIR 1/1211/204/5/2625 Order of Battle of RFC on Somme.1916 July

AIR 1/2129/207/89/1 Notes on Contact Patrol work – Somme area, 1916 July

RFC Records outside AIR 1

Considering the volume of material relating to the RNAS that can be found scattered throughout the ADM series there is surprisingly little specific to the RFC in the War Office records. Individual units frequently refer to aircraft crashes in their vicinity and to aerial fighting going on above them in their War Diaries (in the WO 95 series) but rarely, if ever, do these include details of use to the aviation or family historian.

There are papers relating to applications for Army Commissions by the staff of the Royal Aircraft Factory at Farnborough from 1915 in WO 32/21811–13 which include enlistment papers into the Hampshire Yeomanry.

There is a report on the RFC's participation in the 1912 Army Manoeuvres in WO 33/620 and there are reports from the Committee for Imperial Defence's Air Committee 1912–1914 in CAB 14.

There are high level reports on RFC activities in France and Belgium for parts of 1916 and 1917 in the records of General Headquarters (WO 158/34 and WO 158/35), several sets of papers relating to the RFC in Egypt and Palestine in 1917 and 1918 in WO 158/643–5 and in Mesopotamia in 1915 and 1916 in WO 158/681–6.

First World War prisoners of war

Trenchard pursued a 'forward' policy for his aircraft, encouraging hostile patrols over German occupied territory, whereas the Germans were generally, unless during a major offensive, happier to remain over or behind their own lines. Though parachutes were not worn (more because of their weight and bulk than because senior officers thought they'd encourage men to bale out), a relatively large number of RFC and RNAS officers were forced down behind enemy lines and taken prisoner.

From very early in the war arrangements were made for food parcels to be sent into Germany through the Red Cross. Some of these were provided by the government, but the RFC had its own relief fund. Collections were made and food parcels made up. In spite of the general state of hunger in Germany it seems that the parcels got through regularly and without being interfered with too often. The parcels continued to be allowed through until the Armistice.

There is, unfortunately, no central register in the UK of prisoners of war from the period. For officers, Messrs Cox & Co., the governments' military agents (who arranged pay for officers as one of their duties) compiled a list *British Officers taken prisoner in the various theatres of war between August 1914 and November 1918*. TNA and Imperial War Museum hold copies and your local library should be able to find out whether there is a copy held near you.

There is a list of RFC officers either imprisoned by the enemy or interned by a neutral country in AIR 1/836/204/5/271; another list in AIR 1/892/204/5/696–8 lists prisoners or detainees held in Germany, Turkey and Switzerland in 1916 but there appear to be some names missing.

Soldiers' records at TNA, if they survive, usually contain little information other than recording that a man was a prisoner of war. The responsibility for recording and advising on prisoners came under the International Red Cross and all of their records are now centralized in Switzerland. The address for contact is: Archives Division & Research Service, International Committee of the Red Cross, 19 Avenue de la Paix, Geneva, CH-1202, Switzerland.

Please be aware that records are only available to next of kin and that a research fee is chargeable. You will also need to provide as much information as you can on the prisoner – nationality, name, rank, regiment, date of birth are all useful aids.

If you are able to find a record of which camp your prisoner was held in then you should turn to TNA file WO 161. Part of this file (sections 196–201) contains detailed records of interviews carried out with POWs by the Committee on the Treatment of British Prisoners of War. Every prisoner who came out of Germany was scrupulously interviewed, using a pro-forma questionnaire, about their capture, treatment, their camps, the work they were required to do, sickness, food and recreation. Though the surviving files represent only a fraction of the recorded interviews (something like 5 per cent), they are indexed by name, nationality, camps and a variety of other subjects. It should prove possible, even if you cannot find your man's interview, to locate reports on the camp he was held in. These records have been indexed on-line and can be searched on TNA website.

Captain F B Binney of 12 Squadron RFC was shot down while on a bombing mission over Phalempin during the Battle of Loos on 26 September 1915 and was repatriated in 1918. He gave a long and detailed account of the treatment of prisoners in several camps. He explained that Germany was divided into military districts, the commanding officer of each having considerable independence from Berlin. This meant that standards at camps in different areas could vary considerably. He gives detailed descriptions of food, medical treatment and living conditions at eight camps he was held in, as well as information he had gathered from interviewing other officers and men that he had met. At Schwarmstedt camp 'food was prepared in such a way as to make it as far as possible inedible' and that officers 'carried out the absurd and annoying regulations as to the searching of parcels and tins to the letter' and 'officers in arrest were not allowed to take exercise or to have books, tobacco or food other than biscuits from England'. The Commander of 10th Army Corps, General von Hanisch had supposedly lost his active command while facing British troops and, as a result, 'He will march, surrounded by his staff, and shriek, quivering with rage, at the British officers.' At Holzminden camp he shouted 'that the British were all barbarians and did not deserve to be allowed to live, let alone receive letters'. He was not only unpopular with the British: 'He is known as the "Pig of Hanover" to the German soldier.' For all his complaints about the officers' camps Binney admitted that 'from all accounts soldiers' camps are worse in every way'.

The Foreign Office established a Prisoners of War and Aliens Department in 1914 to deal with all questions relating to these subjects. Though their files are held in FO 383 at TNA, they seem to have been severely weeded. Previously you had to use the Foreign Office card index to identify any files were created on individual prisoners, though many that can be identified this way don't seem to have survived. Fortunately TNA has indexed the surviving files on-line so that they can now be searched using their search engine. The files that do exist cover correspondence with the American diplomats in Berlin and London and with the Dutch, who took over responsibility for our prisoners when America entered the war as our ally in 1917. They also deal with the many voluntary relief organizations that supplied food parcels and with the Red Cross reports.

Lieutenant Douglas Stewart (RFC) has left an incredibly vivid account of his captivity, spread over WO 161/96 and three Foreign Office files, FO 383/266, FO 383/299 and FO 383/302. He describes being shot down:

> When we got over the lines I noticed two German machines under ours so I signalled by means of a red light to the other British machines to go on and take the photographs whilst we tackled these two other machines. We had a regular scrap up and were getting on all right until, quite suddenly, another German machine appeared above us; the three were too much for us and we had our lower right wing shot away and commenced to fall from an altitude of about 6,000 feet. I was acting as observer. The Germans kept on shooting at us until we were near the ground, when they ceased firing. Commander Salmon was an excellent pilot and by a circling movement managed to land the machine on her tail and wing, consequently breaking our fall.

Wounded during the fight Stewart was visited in hospital by his opponents bearing gifts of cigarettes and tobacco. Held at Ingolstadt he often felt hungry, but did attempt an escape, getting out through the wire but being tapped on the shoulder by a German NCO almost immediately. He depended on the RFC Committee food parcels:

> All my parcels used to come through the RFC Committee and these were opened in my presence and nothing was ever taken out. I do not think that I ever lost a parcel . . . and they used to arrive fairly regularly.

Stewart escaped to Holland in April 1917 and, having heard a rumour that the Germans were going to refuse to accept any more parcels, issued a stern warning: 'Our men will certainly suffer acutely, if they do not actually die of starvation if this is carried out.'

There are some lists of known prisoners of war in the AIR 1 series. AIR 1 892/204/5/696–8 give details of POWs in Germany, Turkey and Switzerland respectively. There are some reports of repatriated, or escaped, RAF officers in AIR 1/1206/204/5/2619.

Tying it all together: Lieutenant Christopher Guy RFC

Christopher Godfrey Guy was an officer of 29 Squadron listed as missing on 11 August 1917 and finally noted as having died of wounds suffered after the crash of his aircraft. Using various sources at the National Archive it is possible to reconstruct much of his career and to get some personal information about him.

Lieutenant Guy's service record, compiled in April 1918 when the RFC merged with the RNAS to create the RAF, is in AIR 76/200. It is sadly all too brief. It shows that his father was the Revd F G Guy of 38 Christchurch Road, Bournemouth. Lieutenant Guy transferred to the Royal Flying Corps from the Northamptonshire Regiment on 26 May 1917 and went to 40 Flying Training School. From there he was posted to 29 Squadron on 27 July 1917. On 11 August 1918 he is posted as missing and there is a final note that he 'Died of wounds as a prisoner of war in German hands'.

Because he transferred into the RFC from the Northamptonshire Regiment Lieutenant Guy has an Army Service Record in the WO 374 series. His is WO 374/29863 and, unlike many other files in this series, which have been subsequently 'weeded', it gives a great deal of information. He was born 9 December 1893 and educated at Eton and King's College, Cambridge, where he was a

Abteilung für Gefangenenfürsorge.

VOM ROTHEN KREUZ CENTRAL COMITÉ DER DEUTSCHEN VEREINE

Berlin SW. 11.
Abgeordnetenhaus.

B. 219251

STERBEURKUNDE — ACTE DE DÉCÈS.

'u- und Vornamen ....
'om et prénoms .... — C. G. Guy

)ienstgrad und Nummer der Erkennungsmarke
rade et numéro de matricule .... — engl. Fliegerleutnant

'egiment, Bataillon, Kompagnie ....
'égiment, bataillon, compagnie .... — Royal Flying Corps 1/4 Batt 4

ieburtsort und -tag ....
ieu et date de naissance ....

'ame und Wohnung der Angehörigen.....
'om et adresse de la famille ....

)rt und Tag der Gefangennahme ....
)ù et quand fait prisonnier? .... — 11.8.17. (Flandern)

'erwundung ....
3lessure ....

Tag des Todes ....
Date du décès .... — 12.8.17.

Jrsache des Todes ....
Cause de la mort ....

Grabstätte (Lage, Nummer, Kennzeichnung) . .
Lieu de sépulture (emplacement, numéro, marque distinctive) ....

Ist ein Nachlaß vorhanden und woraus besteht er? ....
Y-a-t-il une succession et de quels objets se compose-t-elle? ....

Bemerkungen (Einzelheiten über die letzten Augenblicke des Verstorbenen und über sein Begräbnis) ....
Observations (Détails sur les derniers moments et l'ensevelissement du défunt) ....

Stempel
Timbre

FELDLAZARETT ... LANDW.-DIV.

Ort und Tag
Lieu et date — O. U., den 19. 1. 1918.

Unterschrift des Ausfertigenden
Signature de l'autorité compétente

Form. 27. 10000. 11. 16

The German Red Cross certificate confirming the death of Lt Guy. (TNA WO 374/29863)

medical student. He enlisted on 5 August 1914 (the day after war was declared), stating that he'd previously been a Sergeant in the Eton Officer Training Corps and had been offered a Commission with them but turned it down as he was leaving school that term.

He had joined the 1st/4th Battalion of the Northamptonshire Regiment and with them had taken part in the fighting at Gallipoli. The 1st /4th Northampton's War Diary (WO 95/4325) covering the campaign doesn't mention Lieutenant Guy by name, but if you read it alongside his WO374/29863 file you can work out what he was doing for his three months at Gallipoli.

On 8 October 1915 he left the battalion due to illness and was taken first to the Base Hospital at Mudros, then on the Hospital Ship *Acquitania* to Southampton. Diagnosed as suffering from typhoid and jaundice he was sent then to the Welsh Hospital at Netley on 27 October. He was still there in December 1915, his jaundice having disappeared but was still infected with typhoid. Lieutenant Colonel Sheen of the Medical Corps stated that he was unlikely to be fit for active service for three months, but that he should be fit for light duties within two.

There is a gap in his record for 1916 but in May 1917 he transferred to the Royal Flying Corps and was sent to 40 Flying Training School. Having passed through the training squadron he was sent to 29 Squadron in the field and joined them. There is a short history of 29 Squadron, written just after the war, in AIR 1/690/21/20/29 which mentions Lieutenant Guy as being a casualty on 11 August 1917. It explains what the squadron was doing for the brief period that he was with them, providing aerial cover for the Battle of Paschendaele that was being fought around Ypres at the time.

The Operations Record Book for No. 29 (Fighter) Squadron mentions Lieutenant Guy only once where he is recorded as 'Missing, subsequently reported DEAD' whilst flying a Nieuport aircraft on 11 August 1917. It was quite possibly his first combat flight.

The combat reports for 29 Squadron are in AIR 1/221/204/5/2637. On 11 August 1917 Captain Oliver records:

> Whilst on OP (offensive patrol) our patrol engaged 3 E.A. (enemy aircraft) Scouts at about 16,000 ft. I dived and fired about 20 rounds into the nearest at 60 ft range. He commenced to spin but recovered at once. I then saw Lt Guy spinning with 2 E.A. (enemy aircraft) on his tail. I attacked the nearest and drove him down.

The Foreign Office made an attempt through the neutral Dutch Legation to find out what had happened to Lieutenant Guy and there is a brief file on him in FO 383/413 which consists of a letter from the British Legation in The Hague, dated 8 January 1918. It states that 'information has been received from the Imperial German Foreign Office that Lieutenant Guy fell in an aerial combat and was buried at Wynedaele, grave No. 12'.

Lieutenant Guy's War Office file (WO 374/29863) contains his German Red Cross death certificate, as well as a translation. Stamped with the seal of the 88th Field Hospital it records that he was unconscious when admitted and that he died without recovering consciousness. His death was as a result of a fracture to the base of the skull, which led to a cerebral haemorrhage. Amongst his effects, which were sent to the Central Office for Effects in Berlin, were a ring with three diamonds, a silver cigarette case, silver match box and silver pencil case.

His body was recovered after the war and he was reburied at Poperinghe New

Military Cemetery and the Commonwealth War Graves Commission website confirms his date of death and gives additional personal information:

> Son of Rev. Frederick Godfrey Guy and Constance Louisa Guy, of 38, Christchurch Rd., Bournemouth. Born Eton. Scholar at Eton College, a Classical Exhibitioner at King's College, Cambridge, and was reading for medicine.

Chapter 3

THE ROYAL NAVAL AIR SERVICE IN THE FIRST WORLD WAR

With the departure of the entire fighting strength of the RFC to France in August 1914 defence of Britain passed, by agreement, to the RNAS, until RFC strength could be increased to the point that they could take the responsibility back. Fortunately for the young RNAS pilots in their poorly armed machines, the expected Zeppelin raids did not materialize until early 1915.

In October 1914 the Eastchurch Squadron commanded by Samson was sent to Dunkirk to support the Royal Naval division that was operating in Belgium. Samson's aircraft patrolled the Channel coast, reconnoitred deep into Belgium and attacked Zeppelin sheds in Germany. Early in 1915 Samson's squadron was sent to Gallipoli where it did sterling service supporting the fleet and the army, and was replaced at Dunkirk by a squadron that grew into a large RNAS presence of several squadrons that became 5 Group RNAS.

During 1915 RNAS aircraft flew and fought at Gallipoli, in East Africa, where they hunted the German cruiser *Konigsberg* in the Rufiji river delta, and patrolled the British coast for German submarines and raiding aircraft. With the start of the Zeppelin attacks they also mounted night patrols but were generally unable to intercept the high-flying airships. On 7 June 1915, warned that Zeppelins were returning from a raid over Britain, RNAS aircraft from Dunkirk patrolled the Belgian coast and Flight Lieutenant Rex Warneford caught Zeppelin LZ 37 as it was descending to its shed near Brussels. Managing to get his aircraft above it he dropped a rack of small bombs, one of which hit and the Zeppelin burst into flames and crashed. Warneford was awarded the Victoria Cross but was killed in a flying accident a few days later.

During 1916, in addition to their air defence responsibilities (taken back by the RFC later in the year), the RNAS formed 3 Wing to launch strategic bombing raids from France against German steelworks in the Rhineland. They had some success but the terrible casualties suffered by the RFC during the later stages of the Somme fighting meant that they were diverted for service under the army. Several RNAS Squadrons flew with the RFC in a fighter role and the RNAS produced their own crop of fighter aces, best known of whom was the Canadian Raymond Collishaw

The officers and men of RNAS depot Mudros, early 1917. (Mrs R J Bone via Mrs R Horrell)

who scored sixty victories against aircraft, as well as shooting down eight observation balloons.

By 1918 RNAS units were serving in the Middle East, in the Aegean, in France and in Italy. They had aircraft operating from a number of ships, including seaplane carriers and planes launched from platforms mounted on battleships and cruisers. The first aircraft carrier as we would recognize it today was close to completion. At 1 April 1918 the RNAS had some 2,900 aircraft and 55,000 officers and men. An order of battle for the RNAS at 1 April 1918 is in Appendix 3.

RNAS Officers

Naval officers joined the RNAS either by transfer from another branch of the Navy, or joined it direct, frequently as members of the Royal Naval Volunteer Reserve. Records of regular officers are in ADM 196. This was the Admiralty *Service Register* which detailed an officer's career as it progressed. These are now available on microfilm at TNA. This series has been partly indexed and the results are on cards held in the Microfilm Reading Room at TNA. The documents were originally large ledgers and, for an officer who served for a long time, the references can be difficult to follow. When the Admiralty clerks ran out of space they'd often cram information into a space on an adjacent page or continue on a page some distance away. Watch out for this, and beware someone else's information being crammed into what appears to be the record for your officer.

There are printed books of 'Disposition of RNAS Officers' throughout the whole of the war between AIR 1/2108 and 2111.

The Fleet Air Arm Museum at Yeovilton holds a series of bound volumes which set out the calculations used to work out RNVR officers' entitlements to a War Gratuity. To do this a brief resume was made of each officer's service. Though there is usually little more information in these ledgers than in the usual service record they can be useful in the event that there are gaps in the service records elsewhere.

A volunteer at Yeovilton has painstakingly gone through every published source of material the museum holds, or can access, to build up a collection of mini-biographies of both officers and men. These are a handy place to begin research as there is an extensive literature on the RNAS and it could save you a lot of time identifying the right books to look in for background information. I spent a lot of time looking for information on a Jesuit priest who had served in the RNAS, with little luck, before I referred to the Museum who gave me several pieces of information from their biography of him which answered some of my questions.

RNAS officers' service records

ADM 273

The main source of information about RNAS Officers services are found in the ADM 273 series at TNA. These consist of thirty bound ledgers which have been indexed in a card form which can be found in the Microfilm Reading Room at TNA. The ledgers themselves are indexed internally. Some men feature in more than one volume, but this is usually shown in the card index, as well as in the volumes themselves.

The ADM 273 record for Woodis Pascal Rogers, shows that he was born on 25 March 1883, son of the Revd C F Rogers of Venton Elwyn, Hayle, in Cornwall. He enlisted as a Royal Naval Reserve Officer on 16 December 1916 for 'Hydrogen Duties'. As all airships used hydrogen for buoyancy he was obviously intended to work with airships but, though there is a space for civil occupation on the record, this has not been completed, so quite how he qualified for this role is unknown. After three weeks at Portsmouth 'N' School on a disciplinary course (i.e. learning how to be an officer) he was posted to RNAS Polegate, an airship station, where he was described as 'a painstaking gas engineer'. He later spent a couple of weeks at the Admiralty Air Department's 'H' section and was then appointed to RNAS Longside at Kingsnorth as 1st Hydrogen Officer, where it was said of him: 'knows his subject and carries out his duties satisfactorily'. He transferred to the RAF on 1 April 1918.

Geoffrey James Pickthall was born on 8 June 1885 and also enlisted as a Royal Naval Reserve Officer on 16 December 1916 for 'Motor Boat Duties'. No next of kin was noted on his record. After a short course in Gunnery at HMS *Excellent* in Portsmouth he was appointed to the yacht *Patricia* and then to RNAS Westgate as an Executive Officer. On 6 October 1917 he was posted to No. 2 Wing, Aegean.

As family historians are aware, sometimes you may find things that are unexpected, and not necessarily pleasant, regarding your ancestor. Medical details are usually given, and several of the ADM 273 records contain summaries of courts martial. The record for Ernest Ferdinand Hast, born 27 January 1887 and enlisted 24 November 1914, manages to combine both. He is shown as being admitted to Haslar Hospital on 12 January 1916 suffering from gonorrhea and being discharged 'refused treatment' two days later. His confidential report dated 4 January 1916 describes him as 'An unsuitable Officer, disloyal to his superiors' while serving at the RNAS Station at Eastbourne. On 25 January 1916 he was tried by court martial, charged with being absent without leave and fraudulent conversion of mess money. He was found guilty and dismissed from the service with disgrace.

Always bear in mind that some men may have more than one record, though this should be recorded on both the index card and on each record itself. Francis Donald

Holden Bremner has records in ADM 273/7 and ADM 273/30, though it is quite hard to say why. The ADM 273/30 record gives few details of his career and, if looked at alone, would suggest that he only saw service at RNAS Dover. In fact the ADM 273/7 record makes it clear that he was also posted to the Aegean, where he was described as 'an excellent scout pilot and has good command of men. Recommended for promotion.' Unfortunately he contracted dysentery and malaria and was returned home where he was judged fit only for ground duties. He became an Executive Officer at RNAS Redcar and was described once again as 'an excellent officer' and a 'V G Officer, G powers of Command'. Unfortunately he was not cleared again to fly during his remaining RNAS service.

Also note that some officers were promoted from the ranks (this should be clear on the record) and you'll need to look for his ADM 188 record for details of his previous service. When looking at the ADM 273 records you'll note that some records are noted as being 'Temporary' and some as 'Continuous'. Temporary service means an enlistment only for the period of the war, in which case there may be further record in ADM 337 (see below). Continuous service means that the officer had previously served as a Royal Naval officer and that records exist elsewhere showing details of his RN service. These records are held in the ADM 196 series and should be looked for there.

ADM 196

Royal Naval officers' service records are held on microfilm at TNA. These too have been indexed in a card form in the Microfilm Reading Room. As well as details of service, and comments of senior officers, they show date and place of birth, name and profession of father, date of marriage and name of wife. They run in parallel with the ADM 273 records for the First World War period and sometimes continue to give information about later service with the RAF.

The record for probably the most famous early RNAS officer, Charles Rumney Samson, shows that he was born on 8 July 1883 at Crumpsall, Manchester, son of Charles L Samson, a solicitor. He went to HMS *Britannia*, the Royal Naval training ship at Dartmouth in September 1897 and was made a midshipman on 15 April 1899 with a Second Class Certificate, having scored 775 marks out of a possible 1,000. He served on HMS *Revenge* and HMS *Victorious*, two battleships, and was described as 'Very steady and promising' and 'Trustworthy and reliable'. He served in the Red Sea as a lieutenant and was awarded the General Service Medal with the Somaliland Clasp for service against the 'Mad Mullah'. He was invalided home suffering from sunstroke and malaria. He is shown as passing his Royal Aero Club certificate in May 1911 and 'Appointed Acting Commander while employed on aviation duties from 1st January 1912: this rank to be effective for purposes of command only as regards Officers employed under him for aviation duties.' As befits his somewhat piratical reputation, the comments of senior officers refer to his 'very good nerve', 'great knowledge of aviation' and to his being 'lacking in tact' and 'wanting in tact'. He was clearly a man who knew how to upset his superiors in pursuit of his goal of building up naval aviation!

The ADM 196 record details his various honours and awards, including the Cross of Chevalier of Legion of Honour for valuable services on the continent 1914–15, his DSO in 1914 and bar to the DSO in 1917 and CMG in the *Air Force Gazette* on 3 June 1919. He was granted a Permanent Commission in the RAF as a Lieutenant Colonel on 1 August 1919. A note inserted into the record notes that Air Commodore Charles Rumney Samson CMG, DSO, AFC of HQ RAF Fighting Area, was removed from the Dangerously Ill List and placed on the Seriously Ill List on 20 April 1929. A final note records that he died February 1931 (*The Times*, 6 Feb. 1931).

RNVR officers' service records

The Royal Naval Volunteer Reserve (RNVR) existed to allow men with an interest in the sea, but who were not part of the Merchant Navy, to join the Royal Navy as a reservist or to join in time of war. On the outbreak of war many men enlisted in the RNVR and were assigned to the RNAS because of their technical knowledge.

The RNVR officers' records consist of microfilmed ledgers and the original Admiralty card index to them can be found in the Microfilm Reading Room at TNA. You'll need to translate the original Admiralty index code to a modern TNA reference in their ADM 337 before seeking out the piece that you want.

The records of service are generally brief, giving full name, date of enlistment and name and address of next of kin. There are usually details of the work the man undertook and of his postings, some of which may expand on details on his ADM 273 service record.

We met Woodis Pascal Rogers and Geoffrey James Pickthall when looking at the ADM 273 records, and in both cases there is a little extra information given in ADM 337. In the case of Rogers it explains that the reason he was sent to 'H' Section at the Air Department was for a training course. In the case of Lieutenant Pickthall it gives his next of kin as his mother, Mrs W M Pickthall c/o Royal Bank of Scotland, Bishopsgate, EC and of The Knightsbridge Hotel, Knightsbridge. It also explains that he was posted to No. 2 Wing, Aegean 'for coding duties'.

One should not expect a great deal from the ADM 337 records, but they can supplement other source material.

Other officer records

As well as the service records in AIR 76 you may find quite comprehensive records for some officers among the group or squadron records. These usually (where they exist) give date of birth, next of kin, full name and details of service with the squadron, but they can sometimes give a great deal of detail.

5 Wing RNAS compiled, as far as can be seen, individual reports on every officer who served with them between 1915 and 31 March 1918. For Flight Sub Lieutenant Charles Philip Oldfield Bartlett, the report runs to nine pages, giving brief details of his service before he joined 5 Wing and then an almost day-by-day schedule and description of his flights and actions. During September 1916 he is noted as having carried out twenty-nine patrols while based at Dover and one fighter patrol over Dunkirk after he joined 5 Wing on 28 September. During October 1916 there was little scope for flying due to the weather but in November there are descriptions of six air raids he took part in on German positions along the Belgian coast, sometimes flying a bomber and sometimes a fighter escort.

On 15 November 1916 he

> took part in a very successful Bomb Raid on Ostende Docks and shipping. On passing Nieuport Piers at 10,000 feet noticed A.A.[anti-aircraft] fire in action along coast at Westende and Middlekerke. On planing down just off Ostende Piers noticed a large blaze a quarter of a mile or so W. of the harbour. Pilot

2nd Lt J W Patterson RAF with his observer Lt Wells of the Sherwood Foresters, attached 34 Squadron, RAF Orvieto, 1918. (With thanks to the Patterson family).

> came down to just over 3,000 feet and dropped bombs in a line N.E. to S.W. over the Atelier de Marine; but could not see results. A chain of incendiary rockets passed very close to machine immediately after dropping bombs, and A.A. guns were very active, but not very close. After dropping bombs pilot circled for some time a mile or so out to sea off the Piers, and noticed a quantity of shipping and several large explosions in the vicinity of the Atelier de Marine.

He is noted as having carried out, during November 1916, one fighter escort flight, one fighter patrol, four bombing raids and eight miscellaneous flights.

Throughout 1917 Bartlett made raids over the Belgian coast and German aerodromes and on 2 July was involved in an aerial fight with a German machine: 'after a number of rounds had been fired the H A went into a vertical dive'. On 4 March 1918 Bartlett left 5 Wing at Dunkirk to join 22 Wing RFC in the field 'and up to the time of leaving was very favourably reported on by his Wing Commander'.

In fact C P O Bartlett was one of the few officers who joined up during the war who was allowed to continue in the new RAF after the war. He continued to serve into the 1920s and, as a result, his service record has not yet been released to the public so the 5 Wing record of service is a valuable point of reference.

Disposition lists of RNAS officers

AIR 1/2108/207/49/1 to AIR 1/2111/207/49/9 contain detailed lists of where officers were based in terms of units they were attached to between 1916 and 1918. These include officers appointed for service at the Admiralty Air Department, giving the section they were attached to, officers attached to squadrons and stations, to overseas expeditionary forces, on board ships and on detached duty with other ministries. There is also a list of officers sick, missing, interned or prisoners of war. Officers are indexed alphabetically so that individuals can be

Squadron Commander C P O Bartlett and members of his squadron at Dunkirk. The photograph was reputedly taken by the Queen of the Belgians. (Liddle Collection with thanks to Richard Davies)

found easily and their postings identified. The lists also serve as a means of identifying RNAS units if you are struggling to identify a set of initials on a service record.

RNAS ratings service records

The service records for any rating who joined the Royal Navy as a regular serviceman between 1873 and 1923 are held in the Registers of Seamen's Services (ADM 188 series) and these include the records for ratings who served in the Royal Naval Air Service. Service details are recorded up to 1928 so you can even trace men who continued to serve for some time after the end of the war. The papers give details about your ancestor's birth, a description of their physical appearance, their occupation and which ship or ships they served on. The good news for people who are unable to visit TNA at Kew is that these records are now available on-line and can be accessed through the 'Records On Line' section of TNA website. The records are searchable on surname and first name, as well as by service number. From 1914 RNAS rating numbers were prefixed with the letter F so if you have a number with this prefix (e.g. F2041) you will know that your ancestor served with the RNAS. Ratings who enlisted prior to June 1914 would not have had this prefix so its absence does not mean that your ancestor was not, necessarily, an RNAS rating.

The service record for George William Blazey (service number F21902) shows that he was born on 10 March 1879 at Wymondham in Norfolk, and that he enlisted on 13 October 1916 as an Acting Airman Mechanic. His civilian occupation was given as carpenter and joiner, so he would have been useful as a 'Rigger', one of the craftsmen who worked on the wooden aircraft frames. After a period at the RNAS station at Cranwell he transferred to Dunkirk and was still there on 31 March 1918 where he was rated as having a very good character and satisfactory ability when he was transferred to the RAF next day. On enlistment he was 5' 7¾" in height, with a 37" chest, brown hair, blue eyes and a fresh complexion.

Albert Edward Turner's record shows that he was born in King's Cross, London, on 28 May 1892. He too enlisted on 13 October 1916 and he too was a woodworker, this time a cabinet-maker and joiner. After his training period he served for six months on HMS *Campania*, a seaplane carrier with the Grand Fleet, and then went to the RNAS station on the Isles of Scilly. He was transferred to the RAF with a very good character and superior ability on 1 April 1918.

One thing to bear in mind with both RNAS ratings and officers is that they were all nominally assigned to a ship for the purposes of being paid. Many of these ships were 'stone frigates' and the man probably never even saw the place he was nominally based, or even ever served at sea. His actual posting should be shown in brackets – Albert Turner is shown as serving in HMS *President II* (Scilly) and HMS *Daedalus* (Scilly Isles). *President* and *Daedalus* are 'stone frigates' and his actual postings were to the Scilly Isles.

With the transfer of RNAS ratings to the RAF their service numbers had a 2 and enough zeros added to the front to make a six digit number. If you know his RNAS number you can adjust it accordingly to find the RAF number that will give you his AIR 79 record for the remainder of his RAF service.

RNAS enlistment papers for men who didn't transfer to the RAF

Most of the enlistment documents for men who transferred to the RAF on its formation in 1918 seem to have been transferred to the new service at the same time and no longer exist. The Fleet Air Arm Museum at Yeovilton does still have the enlistment documents for men who never transferred, either because they had died before 1 April 1918, had transferred out of the RNAS previously or, in many cases, because they were serving with the RNAS armoured car units.

John George Fairclough enlisted for the duration of hostilities on 16 November 1914 as a Petty Officer Mechanic and was posted to Squadron No. 12 of the Armoured Car Division in the Dardanelles (Gallipoli) on 4 August 1915. Unfortunately he contracted dysentery whilst on active service at Gallipoli and died at 19th General Hospital, Alexandria on 10 October 1915.

Herbert Francis Melville enlisted on 23 November 1914 as a Chief Petty Officer Mechanic, but was obviously highly qualified and was quickly promoted to the rank of Sub Lieutenant in the Royal Naval Volunteer Reserve.

Rupert James Anterac enlisted on 14 October 1914 and served briefly at Dunkirk before returning to Britain in October 1915 'to take up a temporary commission in the army'.

The enlistment documents for all three men are held at Yeovilton. All the enlistment papers give a brief physical description, a home address and occupation of the enlisting man.

RNAS/RAF annulled transfers

A small number of RNAS ratings (approximately 2,000) transferred initially to the RAF but subsequently had their transfer annulled. They seem to have been ordinary naval ratings who just happened to be serving at RNAS stations in April 1918, many of them as cooks. Most of their service records should be found in ADM 188 series (see below), but the Fleet Air Arm Museum at Yeovilton holds some paperwork relating to the services of men, whose surnames come in the range M–Z, in a bound ledger entitled 'Statement of Services of RN Ratings in RAF'. Unfortunately the ledger covering surnames A–L seems to have been mislaid at some time.

Files relating to RNAS squadrons and bases are scattered throughout AIR 1 and finding information specific to a unit can be an interesting search.

RNAS records in TNA's AIR 1 series

Records in AIR 1 tend to be grouped under headings such as War Office, Admiralty, HQ Home Forces, HQ RFC or Miscellaneous, which makes searching a little easier using the paper catalogue, but can make searching for them more difficult using the on-line search engine. Searching on the basis of the word 'Naval' is useful and among the 269 records revealed are:

AIR 1/639/17/122/178: Report of service carried out by 3 Naval Squadron, RNAS, with RFC in the field, France. 1917 Feb.–Aug.

AIR 1/659/17/122/599: Various patrol reports 1 Naval Aeroplane Squadron. 1915 Apr.

AIR 1/659/17/122/611: Operations report from 1 Naval Aeroplane Squadron, RNAS. 1915 May 8–12.

AIR 1/1216/204/5/2634: Combat reports: 3 Naval Squadron, 1917 Mar.–1918 Apr.

AIR 1/1218/204/5/2634: Combat reports: 8 RNAS Naval Squadron and 208 RAF Squadron 1916 Nov.–1918 Oct.

AIR 1/1255/204/8/33: Kite Balloon Squadron, Royal Naval Air Service: work summaries 1916 Feb.

AIR 1/1690/204/120/54: Combat reports: Naval Air Squadrons with RFC. 1917 Mar.–May.

AIR 1/1690/204/120/55: Bombing reports: Naval Air Squadrons with RFC. 1917 Apr.–May

Searching on a specific RNAS station is also useful. Great Yarmouth brings up six results including:

AIR 1/187/15/226/4: Great Yarmouth and Felixstowe – daily reports of operations. 1914 Aug. 9–Dec. 31

AIR 1/416/15/243/5: Parts I–III Correspondence to and from Great Yarmouth Air Station 1915 Aug.–1919 Feb.

AIR 1/417/15/243/5: Part IV Correspondence to and from Great Yarmouth Air Station 1915 Dec.–1919 Apr.

However, a search on Yarmouth alone brings up another five, including:

AIR 1/449/15/306/3: Key Plan of RAF Station, Yarmouth 1918;

AIR 1/577/16/15/167: Air raids on England also naval bombardment of Lowestoft and Yarmouth by German Fleet April 1916; and

AIR 1/789/204/4/647: Suitability of a RFC Station at Yarmouth, 1913.

With foreign bases it sometimes pays to search not only on the standard English spelling, but also on the local spelling if known. A search on 'Dunkirk', one of the RNAS's biggest bases, brings up fifty-five results on TNA search engine, among them:

AIR 1/2472: Orders for naval aeroplane and armoured car patrol, and seaplane base at Dunkirk 1914

AIR 1/630/17/122/30: Daily report of 1 Wing, RNAS, Dunkirk. 1916 Jan. 23

AIR 1/634/17/122/98: Report of aerial operations of the Dunkirk Wings, RNAS. 1916 Apr.–July

AIR 1/456/15/312/54: Daily reports of operations: 5 Group (RNAS), Dunkirk 1918

AIR 1/629/17/117/1: Summary of operations: RNAS squadrons, Dunkirk. 1917 July

A search on 'Dunkerque' brings up a further forty-three results, including:

AIR 1/81/15/9/200: RNAS Dunkerque Command – record of hostile machines brought down by. 1915–1918

AIR 1/94/15/9/248: RNAS Dunkerque – daily reports and signals 1917 June–Dec.

AIR 1/75/15/9/173: RNAS Dunkerque Command – Honours and awards gained by officers and men. 1916–1918

AIR 1/108/15/9/288: RNAS Dunkerque – casualties promotions and commissions. 1918 Mar.–Oct.

There are also a number of records for 5 Group RNAS/RAF, which was based at Dunkirk/Dunkerque which have neither name attached so need to be searched for under '5 Group'.

RNAS armoured cars

When Commander Samson went to Dunkirk in October 1914 he enlisted a group of his friends and contacts to serve with him. Among them was the Duke of Westminster who arranged for several of his motor cars (mostly Rolls Royces) to accompany the unit. Such was the lack of reliability of some of the aircraft that they were followed on the ground by a motor car carrying reserve fuel, spare parts and tools. After some clashes with German cavalry patrols the cars were armoured with spare boiler plate and machine guns were mounted on them. A light cannon was mounted on a lorry to give extra firepower. By the time of the First Battle of Ypres in December 1914 the armoured cars were operating in close support of the army and were clearly a success.

Winston Churchill, as First Sea Lord at the Admiralty, was always impressed by new technology and it was decided to form a Royal Navy Armoured Car Division. The HQ was opened at 48 Dover Street near Piccadilly, under the command of Commander F L M Boothby. Men for the division were recruited from all over the Empire. These men were required to sit a rigorous mechanical competence and medical examination. Those who passed were rewarded well, with higher rates of pay and given the rank of Petty Officer (Mechanic). The army was less farsighted; those who were mechanically competent entered the army as a private, with a private's pay.

Originally the Division consisted of fifteen armoured car squadrons, each of which had three sections, and each section was equipped with four armoured cars, one armoured lorry, two supply lorries and eight motor bikes. The Division's establishment was finally increased to twenty squadrons in early 1915. By the time that the RNACD was ready to enter the field, however, its immediate use, scouting in open territory, had ceased to be possible in France because of trench warfare. There were also a number of senior officials at the Admiralty who felt that Churchill was turning the RNAS into his own private service – as well as aircraft, airships and ships it also had armoured cars, armoured trains, kite balloons, searchlights and anti-aircraft guns. When Churchill left office after the Dardanelles debacle the Admiralty began to cut his favourite down to size. Most of its men and equipment were transferred to the army.

When the division was disbanded in the summer of 1915 a number of units had been sent overseas. The 2nd Squadron went to France in March 1915 followed by the 5th, 8th and 15th Squadrons. The 16th, 17th and 18th Squadrons were formed in France. The 3rd and 4th Squadrons were sent to Gallipoli, but after landing a small number of cars, which could not do much good because of the terrain, the squadrons were dispatched to Egypt at the end of August 1915. Other squadrons went to German East Africa, to assist with the capture of German colonies.

One significant unit did remain part of the RNAS, at least until 1918 – the Conservative MP Oliver Locker Lampson established a squadron of cars in early 1915 and agitated to be sent to the Russian front. After some string pulling (he was a friend of the King and had plenty of other high-ranking contacts) the unit was sent to Russia during late 1915 but became icebound at Port Romanov (now Murmansk). It was only in the summer of 1916 that the cars saw action, at first in the Caucasus against Turkey, and later in Romania against Germany and Austria. They took part in the final Russian offensive of the war on 1 July 1917, racing across no man's land to drive the Austrian troops facing them from their trenches, but troops that should have supported them, already totally weary of the war, refused to advance. After the defeat of this final Russian offensive the cars covered the retreat, fighting back superior numbers, but were unable to prevent the final disintegration of the Russian Army. Later that year they played a murky role in an attempted coup d'etat by the Russian General Kornilov and were quietly withdrawn. Reconstituted as an army unit, part of the Motor Machine Gun Corps, in early 1918 they were sent back to Russia, this time to the south, where they fought once again against the Turks until the end of the war.

Another unit was also to remain as part of the Royal Navy. This unit was the 20th Squadron, which was formed as a technical and experimental unit of the RNACD.

This unit assisted the Landships Committee with tank trials and development throughout 1915 and 1916. It can be argued that the tank was actually, in large part, an RNAS development. Certainly it was originally intended that some 'landships' were to have naval crews, but this was abandoned when the navy men learned they'd be on army rates of pay! At the end of the war 20th Squadron had over 600 members.

Armoured car records

In keeping with their chequered history, records relating to RNAS armoured cars and their personnel are scattered. The Fleet Air Arm Museum holds the enlistment papers for the officers and men recruited directly into the RNACD in 1914 and 1915. It is clear that the men were recruited for their knowledge of motors and motoring – Basil Walter Dalrymple (service number F2442) gives his civilian occupation as 'motor journalist' and Dudley Vosper (F2085) is an 'engineer'. The enlistment papers give a brief physical description (Vosper is 5' 5" tall, brown hair, grey eyes with a fresh complexion and a vaccination scar on his left arm).

RNAS service records for the ratings are in ADM 188 series at TNA and can be downloaded from TNA website. Many RNACD men transferred in 1918 to the Army's Machine Gun Corps, where they continued to serve in armoured cars until they were demobbed after the war. It is necessary therefore to look to the Army records for the last part of their war service.

Unit records are scattered between AIR 1 and various Admiralty classes. The Medical Log of the RNACD for March–December 1915 is in ADM 101/332. There are papers relating to casualties at Gallipoli in ADM 1/8424/165, ADM 1/8433/265 and ADM 1/8433/276 and to operations there in AIR 1/668/17/122/722. There is a general report on RNAS armoured car units in East Africa and Russia during 1916 in AIR 1/147/15/64 and specific operations reports on the East Africa cars in AIR 1/658/17/122/667. There are nominal rolls of officers and men of 3, 4, 9, 10, 11 and 12 Squadrons in the Middle East in 1915 in AIR 1/11/15/1/43.

Service records for Locker Lampson's ratings are in ADM 116/1717 and were clearly compiled by the unit itself whilst in Russia, such are the details given of where men served and what they did. There are quite detailed physical descriptions, next of kin names and addresses, also some details of Russian medals they were awarded and of disciplinary offences committed and punishments received. Petty Officer Arthur Cecil Barton, single and aged 26, had previously served with the RNAS armoured cars in France and Belgium but was discharged and joined Locker Lampson's squadron in November 1915. On 17/30 August 1916 (two dates recorded for the same day because the Russians were still using the old Julian calendar) he won the Russian Silver St George Medal in the attack at Haskoi: 'as the Gunner of the 3 pounder trailer under heavy and accurate enemy fire (he) brought his gun into action helping by this action the success of the engagement'. Barton is recorded as having transferred to the Tank Corps OTC (Officer Training Corps) in 1918 so further records should be sought in the War Office files. Petty Officer Frederick Charles Anderson saw service with the unit on the Caucasian Front at Kipriki, August–September 1916, at Dobrudsha November and December 1916 and in Romania. On 16 April 1916 he is noted as having been accidentally killed at Tiraspol base and he was buried next day at the Russian Military Cemetery there. On 18 May 1917 Petty Officer Marcus Devenport Archer 'Had not piped down at 10.45pm contrary to General Orders. Not complying with an order in that when ordered to come forward and speak with the Duty Officer he remained sitting on his bed.' He was sentenced to seven days CB and seven days extra fatigues. Brief details are given at the end of each record as to where each man went on their return to Britain. Most went to the Machine Gun Corps but others are recorded as transferring to minesweeping duties, being discharged sick, remaining in the RNAS and one even as 'Reporting to the Foreign Office'.

Files on Locker Lampson's squadron work in Russia are in AIR 1/662/17/122/667, AIR 1/662/17/122/702 and AIR 1/662/17/122/734. Because it was operating on its own and under the command of the Russian Army, the unit was held to be particularly sensitive and many of its reports were copied to the Foreign Office. These can be found in the Political Files (FO 371) for Russia in 1916, 1917 and early 1918, though you will need to use the card index to locate the reports individually. The Fleet Air Arm Museum at Yeovilton has copies of some of the unit's reports also. A particularly good book on Lampson's unit is *The Czar's British Squadron* by Bryan Perrett and Anthony Lord (William Kimber, 1981). Though Lampson himself had died in the early 1950s, they were able to contact and interview many survivors of the unit.

Armoured trains

For a few months at the end of 1914 and early 1915 the RNAS ran armoured trains in northern France in support of the army. They did take part in some fighting, in particular during the siege of Antwerp in October 1914, where Gunner's Mate T Potter shot down a German observation balloon in six shots from a 4.7" gun and where the train-mounted guns gave useful assistance to the besieged Belgian garrison.

During December 1914 and January and February 1915 the trains shelled German batteries on the coast and inland as far as La Bassee. On 10 January a report reads:

> *Jellicoe* train, under Lieutenant Robinson, supported attack in La Bassee district. After an intense bombardment for 10 minutes infantry attacked and train continued firing slowly. Whilst ranged on bridge on 6-inch shell exploded enemy magazine causing huge flame. Under heavy fire from enemy.

The trains added their firepower to the barrage that preceded the Battle of Neuve Chapelle in March 1915. In particular they were used to attack enemy observation points. On 10 March,

> at 10.49 we obtained a direct hit with 6-inch gun on the tower of Aubers church, which was being used as an observation station. It commanded the whole valley, and it was most important for it to be destroyed . . . At 11.30 a hit from 6-inch gun on Aubers church caused the west wall to fall, and a few minutes later another hit at the foot pinnacle of the tower caused a fire inside.

Shortly after the Battle of Neuve Chapelle the trains were handed over to the army. A printed report on their actions, which includes a nominal roll of officers and men who drove and fought the trains, is in AIR 1/2099/207/21, and there is a brief report on an attack on a train in AIR 1/2629.

Other sources for RNAS records

It is important to remember that the RNAS was part and parcel of the Royal Navy as a whole and that, as well as its own administrative structure, it submitted reports and copied correspondence to the local Senior Naval Officer. This can be useful if you can't trace information on a given squadron or base.

There is some correspondence relating to the RNAS Station at Great Yarmouth in the AIR 1 series but there do seem to be gaps. The ADM 137 series, papers put together by the official historians for the Admiralty after the war, contain a lot more correspondence in the records of the Great Yarmouth Naval Base.

The ADM 137 records can be difficult to find and are not wholly indexed on the National Archive catalogue so you will need to use the paper catalogue and cross-check the old Admiralty references to the newer NA references. Take care not to confuse the old references of HS, HSA and HSB when looking for the NA reference or you'll call up the wrong documents!

For Great Yarmouth the bulk of the material is held in TNA references ADM 137/2234–9. ADM 137/2234 contains information about Zeppelin raids, patrol reports for June–November 1918, Operational Reports June–December 1918, requests for pilots from August 1916 and patrol organization July 1916–May 1917. The remainder hold mainly information relating to the naval base itself, but scattered among them are pieces on airship patrols and air station correspondence which might be worth looking at. As with most official correspondence, ratings are unlikely to be mentioned by name, but there is always a good chance of finding material relating to officers.

RNAS medical records

Medical records specific to individuals are not known to exist, apart from brief notes on the man's service record. One possible source of information is the Surgeons Records held in the ADM 101 class. Most of the records belong to ships, but these do include a number which carried aircraft. HMS *Engadine,* a seaplane carrier (and converted cross-channel ferry) which operated in the North Sea on several attacks on the German mainland, has her surgeon's logs for 1916 and 1917 in ADM 101/413 and ADM 101/426 respectively. The ships surgeons' records are often grouped in alphabetical sequence so frequently, to find a specific ship, you'll have to trace the group in which the log is collected. There are quite a few logs relating to RNAS squadrons and bases indexed individually and as part of the ships group and you can find these quite easily. Eastchurch Flying Schools' log between January and August 1914 is in ADM 101/311, for example, Dunkirk Air Station for 1917 is in ADM 101/441 and Tipnor Kite Balloon Station and Vendome air station are in ADM 101/444.

There is generally a list of patients with a short summary of their period of sickness, a brief diagnosis and details of where they were discharged to. Sometimes there are additional notes, which are incredibly comprehensive. Air Mechanic Arthur George Dew was seen by the doctor attached to the RN Air Service Expeditionary Force at Dunkirk on 1 January 1915 'complaining of shivering and feeling generally ill. Temp 101.8, general pains, a slight cough. Chest examination revealed nothing definite.' Dew was sent to the RN Hospital at Malo, Dunkirk, where he was treated with a purge and 10 grams of aspirin twice a day. 'On the second day he had marked symptoms of a typical lobar pneumonia.' Combined with doses of strychnine and brandy the aspirin treatment worked well.

> I had read of, and been told of excellent results from aspirin in pneumonia but had never used it before. It appeared if given early and in moderately large amounts to be extremely useful.

Not only are illnesses recorded, but also casualties in action. Petty Officer Samuel Mann was employed on an armed boat in the Nieuport Canal when he was struck in the stomach by shrapnel. After first aid treatment he was treated at a Belgian ambulance station and taken to the Anglo-Belgian field hospital at Furnes where he was operated on. Unfortunately he died the same night.

There are some RNAS casualty and sickness admissions in MH 106, representative hospital records (only 5 per cent of the total originally made) saved by the medical historian after the war. These are in MH 106/760, 34 Casualty Clearing Station, May–July 1917; MH 106/987, covering No. 2 General Hospital for February to March 1918; MH 106/1148, covering No. 18 General Hospital, February to March 1918, and ADM 101/106/1335, covering No. 28 General Hospital, April to May 1917.

Ships' logs

A good many aircraft were carried by ships, either seaplane or aircraft carriers or launched by one means or another from existing cruisers and battleships. The logs for all RN ships are in the ADM 53 series. They rarely, if ever, mention individuals by name and are really a log of the ship's position and weather conditions. Halts to launch seaplanes will be recorded, and the launching of aircraft from a platform will usually also be mentioned as it would normally involve a change of course. Logs for the seaplane carrier *Ben-My-Chree*, covering the period March 1915 to July 1916 (when she was sunk by Turkish gunfire) are between ADM 53/35177 and ADM 53/35192. Those for the seaplane carrier *Campania*, April 1915–October 1918, are between ADM 53/36857 and ADM 53/36859. Other ships' logs should be easily traceable using the search facility on TNA website.

The log book for HMS *Empress*, a seaplane carrier that took part in an air raid on the Zeppelin shed at Cuxhaven on the German coast on Christmas Day 1914, gives basic details of her part in the operation as it developed. It gives details of wind speeds and course taken as the ship made its way, along with two other carriers and a flotilla of light cruisers and destroyers, into the Heligoland Bight.

At 06.30 hrs the ship stopped and hoisted out seaplanes Nos. 812, 814 and 815. At 06.49 'Hoisted No. 812 inboard, engine trouble.' At 07.35 a.m. the ship was away at full speed. At 07.56 a.m. a Zeppelin was sighted and at 9.30 a.m. 'Attacked by Zeppelin L6 and three seaplanes. 11 bombs dropped by same.'

No details further details of the action are given, save that 'Lost in 814 seaplane one automatic revolver and leather gear' and 'Lost in seaplane 811 one water bottle and carrier.' Reading the log one would hardly know that *Empress* had been involved in the first modern battle fought by the Royal Navy in the twentieth century.

Chapter 4

THE CREATION OF THE RAF

Having two air services created confusion about their responsibilities and the allocation of scarce resources from early in the war. The Air Board was created in 1916 to coordinate purchasing but the two services continued to operate more or less separately, except where, as in France, a few RNAS squadrons were seconded to work under War Office control. The Royal Air Force was formed on 1 April 1918 from the amalgamation of the RFC and RNAS. The decision to merge the two services was on the advice of General Smuts, Premier of South Africa. Smuts had been asked to look into the aerial defences of the UK after two daytime bombing raids in the summer of 1917, which killed and injured hundreds of Londoners. He was not impressed by the breakdown in communications and division of responsibilities between the Army and Navy that twice allowed the Germans to get through the defences and escape relatively unscathed, and recommended the creation of one air service. Many said that the 1 April date was significant, and the new service was not anticipated to survive long once the war was won. The minutes of Smuts Committee on the new Air Ministry and amalgamation of air services are in CAB 21/21.

In the final months of the First World War the RAF helped to contain the German offensives that threatened to sweep through to the channel ports and later took part in the great advance that drove them back across France. New tactics evolved which used a potent combination of ground attack aircraft, tanks, artillery and infantry to smash through the German defences. RAF fighters kept German fighters at bay and bombers attacked their bases and industries in the Ruhr. On the day that the Armistice was signed a squadron of RAF heavy bombers was poised to make the first air raid on Berlin.

On 11 November 1918 the RAF comprised 188 squadrons, 22,000 aircraft and nearly 300,000 officers and men. It was rapidly cut to just thirty-three squadrons. Hugh Trenchard, who had commanded the RFC in France for most of the war, was appointed to head it.

RAF records for all units up until about 1921 continue in the AIR 1 series. Service records for officers who left service before the end of 1919 are in AIR 76 and airmen who left service before the end of 1922 are in AIR 79.

Some interesting files on the formation of the RAF and the effect it had on the component services are:

AIR 1/533/16/12/112: Formation of Royal Air Force. Administration of RFC during transitional period;

AIR 1/618/16/15/341: Order in Council: transfer of RFC and RNAS into Royal Air Force;

AIR 1/683/21/13/2231: Orders of formation of Royal Air Force and conditions of pay, commissions, billeting, attachments, etc;

AIR 1/912/204/5/844: Correspondence, orders and pamphlets: Formation of Royal Air Force;

AIR 1/2405/303/4/9: Air Force Memo 3: formation of the Royal Air Force;

WO 32/9289: Transfer of army personnel to Royal Air Force;

WO 32/9291: Transfer of responsibility for payment of army personnel, Royal Flying Corps to Royal Air Force.

Records of the RFC, RNAS and RAF in other theatres

The main theatre of the war was France and Belgium and this is where the majority of RFC, RNAS and RAF squadrons operated, but they served virtually everywhere else, with units in Italy, Egypt, Palestine, Mesopotamia, East Africa, Greece, Romania, India and Russia. The service record should say that a relative served overseas and clues to look out for, as well as the obvious naming of a country, include the letters 'EF' (Expeditionary Force) among initials – i.e. EEF is Egyptian Expeditionary Force, MEF is Mediterranean Expeditionary Force and NREF, North Russia Expeditionary Force.

Try searching TNA catalogue using the country as a keyword, but also cast around. Using Greece as a keyword produces only three AIR 1 records, but Aegean produces sixteen! Also be aware that sometimes spellings change: Rumania as a keyword brings up one record but Roumania (the more common spelling at the time) brings up two different files. You'll also find nothing under 'Kenya' but sixteen files under the slightly wider East Africa. What is now Iraq was then Mesopotamia.

As an example, there are numerous files relating to the early RFC, RNAS and RAF units in Mesopotamia in TNA, most of them in the AIR 1 series. AIR 1/140/15/40/306 gives details of the moves of machines and personnel of 30 Squadron to Mesopotamia 1915 to 1917. AIR 1/504/16/3/23 is a report on Aviation and flying operations in Mesopotamia 1915 to June 1916. AIR 1/121/15/40/110 contains RFC summaries of operations for the first half of 1916. AIR 1/648/17/122/386 covers operations and administration of RNAS in Mesopotamia and AIR 1/648/17/122/392 is a report of monthly RNAS operations, Mesopotamia.

AIR 1/2263/209/61/1–5 are the War Diaries of the initial RFC unit in Mesopotamia. AIR 1/466/15/312/155 are the weekly RAF summary of operations, Mesopotamia in 1918.

The Women's Royal Air Force

On 1 April 1918 The Women's Royal Air Force (WRAF) was formed from those women already serving in air units of Women's Royal Naval Service, Women's Auxiliary Army Corps, the Voluntary Aid Detachment and the Women's Legion. Women took over many traditionally male occupations such as driving, working on engines and rigging aircraft, as well as more 'traditional' roles of administration, cooking and cleaning. Service was, at first, strictly in Great Britain, and there were two classes of servicewoman, mobile and immobile. Immobile women contracted to serve only at their local RAF establishment whereas mobile, as the name implies, could be ordered to serve anywhere. After the Armistice some 500 women were allowed to serve abroad, either in France and Belgium, or with the Army of the Rhine. At its height, the WRAF had 24,659 members but budget cuts following the end of the war caused the force to be disbanded in 1920.

The WRAF was not without controversy. The first Commander was a man, Lieutenant Colonel Bersey, with Lady Gertrude Crawford as his Chief Superintendent. Lady Gertrude soon realized that she was just a figurehead and resigned, and Lady Violet Douglas-Pennant took over as Commandant. She soon began to suspect that the RAF weren't taking the WRAF seriously. There were problems in some of the camps over the treatment of women by male RAF senior officers. After a critical report on the organization by Lady Margaret Rhondda, Douglas-Pennant was dismissed. In September 1918, Mrs Helen Gwynne-Vaughan, who had gained a reputation as an efficient administrator in the Women's Auxiliary Army Corps, took charge of the WRAF. Gwynne-Vaughan was a great success as commander and General Sir Sefton Brancker argued that 'the WRAF was the best disciplined and best turned-out women's organization in the country'.

Lady Violet Douglas-Pennant did not take her dismissal lightly and demanded an official inquiry. It was heard by the House of Lords but Lady Violet made false accusations against some of the witnesses and was even sued for libel by two of them. She later wrote an interesting book *Under the Searchlight*, which gives her side of the story and some details of life in the WRAF.

The Women's Royal Air Force: officers records

No records for the officers who served in the WRAF during the First World War are thought to survive. The RAF List, which was introduced in April 1918, does not contain mention of the WRAF officers. Some senior officers are named in Lady Douglas-Pennant's book.

The Women's Royal Air Force: other ranks

Unlike for the officers, personal records do exist for many of the other ranks of the WRAF, though what there is is a bit of a hotch-potch. They are held on microfilm at TNA in their AIR 80 series – alphabetically, so you don't need to know a service number to locate your relative. One thing to note is that if the servicewoman enlisted under her maiden name then subsequently married the record will be held under her married surname.

For some women all that survives is their certificate of discharge on demobilization. This gives a brief physical description, their service number, name Air Force Trade, date and place of enrolment, date and place of demobilization and brief (usually one or two words) description of their work and personal character.

The form of enrolment in the Women's Royal Air Force gives similar information to the discharge certificate but includes information on the woman's health, next of kin and category of work (A, B, C, or D). These categories represented:

Category A: clerk, shorthand writer, typist

Category B: cleaner, cook, waitress, laundress, domestic worker

Category C: Chauffeuse, photographer, fitter, tinsmith, metal worker, rigger, wireless mechanic, wireless operator, carpenter, painter

Category D: storekeeper, tailoress, shoemaker, sailmaker, motor cyclist.

Sometimes there is a statement of service attached which tells you where the woman was based, but on others you'll need to check the stamp of the approving officer for the form, which should tell you where she was based.

The third record type, which can cause confusion, is the Casualty Form – Active Service. These don't appear to have been used (as they normally should have been) as a record of sickness or injury, but rather as a short service record kept and updated by the base at which the servicewoman was stationed – in fact they're quite useful for this as they had preprinted spaces for just about everything the recording officer would require. Catherine Josephine Clarke's Casualty Form shows her to have enlisted on 21 September 1917 into the Women's Auxiliary Army Corps, aged 28 years. She enlisted for the duration of the war as a general clerk and was posted to the Central Flying School at Upavon in Wiltshire. She was admitted to the Isolation Hospital at Tidworth on 9 January 1918 and released after 12 days. In March 1918 her name was changed through marriage to Venter, and she transferred to the WRAF on 1 April 1918.

Though most records are merely lists of postings one does occasionally come across something more unusual. Vera Lillian Bates was a storewoman at the Armament School in Uxbridge when she was discharged on disciplinary grounds in February 1919. Closer examination of her record reveals that 'disciplinary' had been written in over the original reason which was 'compassionate'.

Apart from the other ranks records, there are not many records at TNA relating to the WRAF. There is a brief history, written in the 1930s in AIR 1/681/21/13/2212 and a file on the work done by the WRAF in 5 Group (Dunkirk) in AIR 1/106/15/9/284. The constitution and Regulations for the WRAF are in AIR 1/619/16/15/347 and in AIR 1/2087/207/7/14.

Tidying up after the war

With the Armistice the British Army advanced into Germany to occupy the Rhineland. Their part in the Occupation lasted from 6 December 1918, when 7 Squadron with their RE8 aircraft, arrived at Elsenhorn near Cologne, until 21 September 1919, when most squadrons departed. Only one flight of 12 Squadron stayed behind until it was withdrawn in early 1923. Some twenty squadrons spent some time in Germany – Nos. 7, 9, 11, 12, 18, 22, 25, 29, 43, 48, 49, 59, 62, 70, 79, 84, 149, 206, 207 and 208 – all of which were based in the Cologne area.

Apart from specific squadron records there are several files relating to the Occupation, including:

AIR 1/1025/204/5/1411: Move of personnel to Rhine Area;

AIR 1/1015/204/5/1344: Move of 9 Aircraft Park to Rhine;

AIR 1/1152/204/5/2403: Moves of Units as Army of Occupation and of RAF HQ on the Rhine;

AIR 1/1594/204/83/22: Move to Rhine instructions;

AIR 1/1039/204/5/1469: Re-organization of RAF for Army of Occupation in Germany; AIR 1/1114/204/5/1921: Lists showing location of RAF Units, Army of Occupation and ultimate disposal.

RAF also operated an air mail service to Germany which contributed greatly to keeping up the morale of the troops serving in the Rhine Army. There is a resume of its work in AIR 1/1115/204/5/1953, 'Work summary and returns from Aerial Post between England, France and Germany', and some details of their work in AIR 1/1114/204/5/1937: Postal arrangements, including aerial posts: Army of Occupation.

The RAF in North Russia 1918–19

Even before the Armistice with Germany the RAF became involved in a series of operations in Russia against Lenin's Bolshevik Government. In August 1918 a full-scale invasion, including the seaplane carrier HMS *Nairana* was launched to capture Archangel and a small RAF contingent landed to train White Russian pilots. The winter freeze meant that troops couldn't be evacuated and both sides settled down to wait for the spring.

With the approach of spring 1919 reinforcements were sent to both Archangel (Elope Force) and Murmansk (Syren Force). They were to hold the line, allowing the troops who had been in Russia all winter to evacuate. They would also train White Russian troops and aviators into a new army to continue the war against the Reds.

HMS *Nairana* reached Murmansk on 28 March 1919 and, after testing her seaplanes, began to send them south to the front line near Lake Onega. The country was thickly forested but had many small lakes and rivers that made seaplanes ideal aircraft to operate in the area. Other troops and aircraft advanced inland from Archangel down the River Dvina, hoping to link up with White Russian forces advancing out of Siberia.

Nairana's seaplanes established a base on the shores of Lake Onega and took part

The temporary RAF seaplane base at Medvejya Gora, Russia, 1919. (Mrs R J Bone via Mrs R Horrell)

Relaxing – RAF Lieutenant Guy Blampied prepares for a swim in Lake Onega, Russia, in the summer of 1919. His Fairey IIIc aircraft is behind him. (Guy Blampied)

in some quite sharp fighting in support of Royal Naval motor boats in action against Bolshevik shipping. On 6 June 1919 they erected their first aircraft, which they had brought, partially dismantled, from Murmansk by train, and undertook their first reconnaissance that day. Over the next four months *Nairana*'s seaplanes (known as 'Duck Flight') accompanied later by a mixed force of land-based aircraft helped the army push south down the railway line to what was hoped would be a strong defensive line to be held by the White Russians while British forces were withdrawn.

The RAF units sent to Archangel on HMS *Pegasus* and the brand-new aircraft carrier HMS *Argus* also advanced inland and fought a series of actions against the Bolsheviks until both forces were withdrawn in September 1919.

South Russia

During the summer of 1919 other RAF units were sent to South Russia to assist White Russian forces operating there. Makeshift seaplane carriers operated on the Caspian Sea and flew raids against Bolshevik shipping there and in the River Volga. Other units of land-based aircraft operated from temporary airfields, both training the Whites and flying their own operations. Some units, including 47 Squadron, commanded by former RNAS officer Ray Collishaw, remained in Russia until 1920.

Did my ancestor go to Russia?

Service records in AIR 79 or AIR 76 may refer to 'RAF Elope' for Archangel and 'RAF Syren' for Murmansk, as well as just NREF (North Russia Expeditionary Force) or Kem, Medvejya Gora (places on the Murmansk front where the RAF were based) or Bereznik (on the Dvina/Archangel front) or SREF (South Russia Expeditionary Force). A higher proportion of men who went to Russia seem to have stayed on in the RAF afterwards than normal so if you think your ancestor went to Russia but can't find a service record this might be the reason. There is a nominal roll of officers and other ranks attached to RAF Syren in AIR 1/1768/204/143/2. Nominal rolls for South Russia are in AIR 1/1666/204/99/12 (Other ranks) and AIR 1/1666/204/99/13 (Officers). AIR 1/1960/204/260/40 contains Nominal rolls of British and Russian personnel, Headquarters RAF South Russia.

Finding the records

In order to trace all the relevant records for the RAF in Russia during the period it isn't just a question of using TNA search engine to look for Russia, though this is a good starting point. It is useful to search under Syren and/or Elope, as well as Murmansk and Archangel, when looking for relevant files. Most you will find in the AIR 1 files, but there is some information in AIR 2. As the operations were closely linked to both Army and Naval units it is also worth checking for anything in the WO (War Office, who ran the Army) and ADM (Admiralty) files.

There are fifty-nine AIR 1 files that come up under a search for 'Russia', though of course not all are related to 'Syren' or 'Elope'. A search on 'Syren' alone brings up one AIR 1 file that does not come up on the 'Russia' search (AIR 1/450/15/312/10), which contains pilots' and observers' reports. There are also nine WO files that are produced by this search, and one of them (WO 106/1151) contains a list of the RAF officers who first went to North Russia in 1918.

The quality of the records for both RAF Syren and RAF Elope are variable and do not seem to have been collated with the same care as most AIR 1 records. This does mean that often what you'll find on the files are the actual notes so you can actually handle documents created by your relative – the disadvantage is that you may not be able to decipher their handwriting.

Somaliland

For over twenty years the government of the Somaliland Protectorate had been engaged in a guerrilla war with Mohammed bin Abdullah Hassan, known to the British as the 'Mad Mullah'. Though often defeated he had always resurfaced and, on occasion, inflicted some terrible defeats on the British. In 1913 he had inflicted a serious defeat on the Camel Corps and, while attention had been turned towards the war in Europe, he'd fought a series of actions against the small British garrison and was obviously still a force to be reckoned with.

In 1914 the use of either aircraft or airships against him had been considered (the files are in AIR1/2526 and AIR 1/625/17/6) and in 1919 Trenchard agreed to send a small force of DH 9 bombers (known as Force Z) to help pacify the unruly protectorate. It was hoped that this use of technology would save money at a time of immediate post-war retrenchment. In late October 1919 an advanced party left London to set up bases and find suitable air strips. The main party followed and reached Somaliland at the end of December 1919 and the erection of the crated aircraft began on New Year's Day. In a campaign lasting just twenty-one days the DH 9s, assisted by the local garrison of King's African Rifles and Camel Corps, bombed and machine-gunned the Mullah's forts, followers and flocks. The Mullah fled from his fortress, relentlessly pursued by the Camel Corps and the RAF. His power in the protectorate was broken in just three weeks and he later died (possibly of the flu).

The value of aircraft in dealing with 'troublesome natives' cheaply had apparently been confirmed and Trenchard was able to make the argument that the RAF should be given control of Iraq, a role that it formally adopted on 1 October 1922.

The RAF contingent in Somaliland consisted of just thirty-six officers and 183 airmen but they set a high standard for their fellows. The medal entitlement was for the Africa General Service Medal, with a 'Somaliland' clasp, and the list of recipients is in AIR 2/2270, with recommendations in AIR 2/204. Reports and communiqués are in AIR 1/23/15/1/116, AIR 1/36/15/1/238 and AIR

1/2417/303/36/1. The unit war diary is in AIR 5/1309 and other papers, including operational orders, reports on medical arrangements and observers' reports can be found between AIR 5/1310 and AIR 5/1315.

Mesopotamia

With Turkey's surrender, responsibility for Mesopotamia passed to Britain under a League of Nations mandate. Britain was responsible for creating the new state and guiding it to full independence, but this was not necessarily how the locals saw it. There was considerable tribal unrest, rioting in some of the cities and, in the north, Sheikh Mahmud's rebellion pushed for Kurdish autonomy. Local garrisons were besieged, trains derailed, telegraph lines cut. Reinforcements were rushed from India. The two RAF squadrons permanently in what was now being called Iraq flew constant missions attacking rebels with bombs and machine guns. The rebellion was crushed by a combination of force and an offer to place an Arab prince, Faysal, on the throne. A constitutional monarchy was established with a freely elected parliament, but with British 'advice' on foreign, military and judicial affairs.

Among files on the rebellion of 1919/1920 are:

AIR 1/2326/223/53/4: Report by Mesopotamia Commission on operations of war in that country;

AIR 1/2357/226/5/18: War diary and resume of Mesopotamia operations;

AIR 1/22/15/1/114: Resume and diary of operations of Mesopotamia Wing, RAF;

AIR 1/431/15/260/18: Armoured car reconnaissance reports, Iraq.

Chapter 5

THE RAF BETWEEN THE WARS

The RAF Reserve

The RAF Reserve was created in 1922 and many former officers, who had applied for permanent commissions in 1919 but been rejected, saw it as a way of keeping up their interest in flying at the government's expense. Though their service would have extended well beyond the 1922 cut-off date for release of service records, the good news for family historians is that the weeding process when the service records were transferred to TNA does not seem to have been at all thorough. Though for many, alongside their First World War record, you will simply find a page marked 'Transferred to the Reserve Ledger', showing that their Reserve record has been moved, in perhaps as many as 30 per cent of cases the record is still there in AIR 76. Some of these continue well into the 1930s!

Percy George Clarabut, formerly a Second Lieutenant in the East Kent Regiment (his Army record is in WO 339/100745), transferred to the RAF in May 1918. He had not completed his training by 11 November 1918 so never went overseas with the RAF. He was transferred to the Unemployed List in April 1919. He had obviously enjoyed flying, however, as in March 1923 the Aviation Candidates Medical Board (ACMB) recorded that he was 'Fit for flying duties as pilot'. He attended a requalifying course at the De Havilland Civilian Flying School for a month in August 1923 and was reported as 'Has been satisfactory in every respect. His cc [cross country] flight to Manston was completed in misty weather quite satisfactorily.' There are reports on his annual training courses up to 1928 ('good sound pilot') and notes that he continued in the Reserve until 1934. A note with his records says 'Transferred to Reserve Ledger' so perhaps his later reports were transferred and the early ones missed?

The record for Ronald Cory Berlyn covers the whole of his Reserve service from his requalifying course in 1924, through his training to be a Flying Instructor at the Central Flying School in 1926 – 'a good pilot and above average as an instructor' and

> Is a steady and consistent instrument pilot who has no special difficulties in learning. Has quite good knowledge of method of instruction and is considered capable of instructing reserve pupils in instrument flying.

There are various addresses recorded for Ronald Berlyn, all in the Midlands, and his marriage to Rona Mary Hallam at Birmingham South Registry Office is recorded on 6 October 1927. The record even notes the number of hours that he flew each month between September 1926 and June 1931. He transferred to Class C Reserve in June 1935.

If you know that your relative continued to fly with the Reserve (or to serve in an administrative capacity with them, as some officers did) then don't assume there isn't a record in the public domain – check AIR 76!

Operations Record Books (ORBs) – standardization of RAF record keeping

Once you have gone through the service record and identified where your relative served, the Operations Record Books (ORBs) are the next logical step for investigation. They are probably the key documents for finding out what the RAF unit your relative was posted to was doing on a day-to-day basis, particularly during wartime or in time of crisis.

The ORBs, in the form that we generally know them, were introduced at the end of 1926 and were designed to bring a degree of conformity to the way the RAF kept its unit histories. They were based on the Army's War Diaries, with the exception that the ORBs were to be completed during peacetime as well, though the degree of detail in peacetime is usually sparse – frequently whole weeks are summarized by 'Station work continued as normal'. The level of detail also depends to some extent on the personality of the officer who was tasked with writing up the ORB; some men were very conscientious, others far less so. So what are ORBs and what do they contain?

Operations Record Books, comprising both 'Summary of Events' forms (Form 540) and 'Detail of Work Carried Out' forms (Form 541), together with their appendices (usually operational orders, miscellaneous reports and telegraphed messages) are a record of daily events kept by all units of the Royal Air Force. If you are lucky, an ORB will contain lists of names (on duty rotas, for example, or on transfers of personnel elsewhere) and some contain photographs. The information recorded can vary, often depending on circumstances at the time the ORB was written – some squadron and station records from the Battle of Britain period are sketchy. Quite simply there was far too much going on for time to be spent on the luxury of writing up the ORB. There are also occasional gaps – 605 Squadron was overrun by the Japanese on Java in 1942 and their ORB for their period in the Far East must be presumed lost in the rout. During quiet periods, particularly for some of the smaller training or depot units, all you might get is a note 'nothing of interest to report'.

Though ORBs were kept at command, group and wing level (in AIR 24, AIR 25 and AIR 26 respectively), the ones most likely to be of interest in researching a relative are the ones for squadrons, stations and miscellaneous units (AIR 27, AIR 28 and AIR 29 respectively). Quite often you'll be able to refer to ORBs for both the squadron and the station it was based at – these complement each other, the squadron ORB being usually concerned with events in the air, the station one also covers these (and will tell you about some of the other units that shared the station) and also more about events on the ground..

When searching for group ORBs on TNA website search engine you'll need to be aware that the name of the group frequently reflected its role, so that 1 Group cannot be found by a simple search on that term. No. 1 Group is actually referred to as NO. 1 (BOMBER) GROUP and its ORBs can be traced using this reference. If you're not sure of the exact title but have the number of the group the good news is that until 1946 the records are in numerical order so you can browse through until you find the ones you want.

You should also be aware that early ORBs, usually covering the 1920s and 1930s, are not easy to find using TNA search engine, particularly for squadrons (in AIR 27). Simple searches on '5' (hoping to find 5 Squadron) and looking in AIR 27, appears to suggest that ORBs begin in 1949. You'll notice, however, that there is a subsection 'Subseries within AIR 27'. If you click on the underlined NO. 5 SQUADRON and then on the 'Browse from here' box you will be presented with a further list of 5 Squadron ORBs from July 1913 to July 1947 (AIR 27/63 to AIR 27/66).

There are a couple of things you need to be aware of when consulting an ORB. They were introduced in their generally accepted form in 1926, but some units felt it worthwhile to write a brief history back as far as the unit records available would allow. This means that you will see some ORBs in AIR 28 noted as starting as back as far as 1913 and 1915 (Calshot AIR 28/120, Hornchurch AIR 28/384, Lahore AIR 28/440 and Marham AIR 28/516) but strictly speaking these are not contemporary reports so should be treated with a degree of caution. For First World War squadron record books you need to look back into the AIR 1 series. Another related point is that it is clear that some ORBs that you might think were reasonably contemporary with the events they describe are, on further examination, found not to be. The ORB for 5 Squadron (AIR 27/63) which claims to begin in 1913, and which begins to get detailed in 1925, is actually written in an ORB with a print date (i.e. when the blank ORB was produced) of 10 September 1937!

Many of the station and squadron ORBs are held on microfilm and can be accessed in TNA Microfilm Reading Room, where printing facilities are also available. Others come in bound ledgers (which I always prefer to read) and you can read them at your desk. Unfortunately there is no way of knowing, if ordering remotely from home, which ORBs are on microfilm and which need to be ordered in advance – I usually play safe when time is precious and only try and order ORBs whilst at TNA itself – in this case the system alerts you that the documents are filmed so you can use your orders sparingly and only order material that needs to be produced from TNA's vaults.

We'll look in some detail at a selection of each type of ORB later on.

The RAF List

The RAF List can be an invaluable source for tracing officers' careers and for finding out about squadrons and stations in the inter-war period.

Published monthly by the Air Ministry, its purpose was to show promotions and postings of officers (other ranks aren't mentioned, with the exception of some senior non-commissioned officers). Even without an officer's service record you can use the volumes to track his promotions and postings, and it is easy to use as it is indexed.

By locating his postings you'll be able to see where he was based, who his fellow officers were and, until 1939, the type of aircraft that his squadron flew. The List is also valuable for checking abbreviations on the service record because if you know where he was at a given time you can match the abbreviation with the posting. It also lists Royal Navy officers serving in the Fleet Air Arm and gives details of the aircraft carriers on which they were serving.

TNA holds a comprehensive run of Air Force Lists in its Microfilm Reading Room and many large libraries also have them tucked away in their reference sections, though you'll probably have to ask for them to be retrieved from their stack. Individual copies should be orderable from the inter-library loans system.

The RAF abroad between the wars

Hugh Trenchard, as Chief of Air Staff, was determined to maintain the independence of his fledgling service and saw great opportunities for it in policing the Empire. He also made policy that as many officers and men as possible were to serve overseas.

Iraq

The RAF's biggest commitment in the 1920s was in Iraq, as a result of Trenchard's deliberate policy of keeping the service active abroad, and because he claimed that pacification from the air was cheaper by far than using a huge commitment of ground forces. Ever keen to save money, the British Government was happy to try this novel experiment.

To support the Iraqi Government in its path to independence 'as soon as possible', it was still necessary to have a British garrison and, following the success of operations in Somaliland, Trenchard proposed to use 'Air Control' as the best means of holding down the country. Air Vice Marshal Sir John Salmond became commander of all British forces in the country in October 1922.

By October 1925 the Iraq garrison was eight RAF squadrons (92 aircraft); three armoured car companies (63 vehicles); one British (2nd Beds & Herts Regt) and

A Hinadai troop carrier aircraft, Iraq, c.1927. The RAF in Iraq pioneered the rapid movement of troops by air using such aircraft. (Author's collection)

three Indian battalions and support troops (totalling 3,000 men), 4,700 Iraqi levies and an Iraqi Army of 7,100. This was a considerable reduction on the over 125,000 men required in 1921. There were further rebellions in the north led by Sheikh Mahmud but the RAF developed a policy of bombing towns and villages identified as being in rebellion and attacking the flocks and herds of rebel tribes. One of the squadron leaders responsible for developing the practice was Arthur Harris, later to become famous as 'Bomber' Harris in the Second World War. Between 1925 and 1931, when Sheikh Mahmud was finally captured, barely a month went by without the RAF being in action. During periods of serious trouble they were in action on a daily basis, often working in support of armoured car units and the growing Iraqi Army. On one occasion a loudspeaker was mounted in an aircraft that flew over hostile villages warning of the consequences of further rebellion, much to the consternation of the villagers.

In September 1932 Iraq obtained full independence and there was a slow withdrawal of RAF and other British units from the country, though a presence always remained.

RAF Squadrons in Mesopotamia and Iraq

30 Squadron	mid 1915 to Second World War
63 Squadron	August 1917 to February 1920
72 Squadron	March 1918 to February 1919
6 Squadron	April 1919 to Second World War
55 Squadron	August 1920 to Second World War
70 Squadron	1922 to Second World War
84 Squadron	August 1920 to Second World War
1 Squadron	April 1921 to November 1926
8 Squadron	1921 to February 1927
45 Squadron	March 1922 to Second World War

Records of the RAF in Iraq

There are over 350 files relating to Iraq in the AIR series including the following. Regular reports on the Iraq Command in AIR 10 cover 1922 to 1934. AIR 5/724 are the Despatches of the Air Officer Commanding, Iraq for 1926. AIR 5/1253 to AIR 5/1255 are reports on operations in Iraq 1918 to 1937. AIR 5/1287 to AIR 5/1294 are Iraq Command monthly operations summaries 1921 to 1939. AIR 5/256 are operational reports 1921 to 1926. AIR 20/187 covers RAF operations 1929 to 1932. AIR 20/746 covers Operations on Iraq's northern frontier 1924 to 1925. AIR 23/249 to AIR 23/253 are resumes of operations in Iraq 1928 to 1932.

AIR 27 contains Operation Record Books for most of the squadrons based in Iraq. AIR 29/833 is the Operations Record Book for RAF Hinaidi Central Supply Depot for Iraq, 1922 to 1943.

There are hundreds of Iraq Command files from the 1920s and 1930s, dealing with (among others) local operations, tribal politics, reports on specific localities, intelligence, arms trafficking, suspicious characters, Kurdish Nationalist Movements and possible Turkish intervention. Iraq Air Headquarters War Diaries between January 1923 and December 1930 are on a monthly basis between AIR 23/457 and AIR 23/541.

Squadron Leader C H Keith spent three years in Iraq with 70 Squadron and as commander of 6 Squadron, and wrote an excellent book based on his letters *Flying Years* (The Aviation Book Club, 1936) which describes his daily work and the nature of the country.

The RAF in India

The first RFC unit to be sent to India was part of the newly formed 31 Squadron in September 1915. By the summer of 1916 the whole of 31 Squadron was based at Muree and was employed supporting the army on the North West Frontier. In 1917 part of 31 Squadron was separated off to form the nucleus of 114 Squadron and both squadrons operated in support of the army along the frontier and in quelling riots. In 1919 an invasion of India by Afghanistan gave the RAF the opportunity to mount a long-distance bombing raid against targets in Kabul using a Handley Page

1600 | 1601 | 1602 | 1603

DEPÔTS.

DEPÔTS.

R.A.F. DEPÔT, MIDDLE EAST.
Middle East.
Aboukir.

Group Capt.
Kennedy G. Brooke, *C.M.G.* 24Sept.26
Wing Cdr.
Frederick W. Stent, *M.C.*[e] 17Oct.26
Sqdn. Ldrs.
Patrick A. O. Leask [e] 17Oct.26
William S Caster *M.C* 19Jan.27
Clair St. Noble [e] 17Oct.26
Flight Lieuts.
John F. A. Day, *A.F.C.* [e] 6Nov.26
Alexander J. Long 6Nov.25
Malcolm F. Browne [e] 17Oct.25
John G. S Candy, *D.F.C.*[e] 2Jan.27
Garrett M. F. O'Brien, *D.S.C.* [e] .. 1Sept.27
Cecil F. Chinery [e] 7Aug.27
Cyril Chapman, *D.S.C.* 9Feb.27
Hubert P. G. Leigh [e] 22Sept.25
Hubert J. Adkins [s] 16Jan.25
Flg. Offs.
Norbray Liddall [e] 18Oct.24
James Rodger, *D.S.M.* [e] 14Jan.27
William Gill [e] 17Oct.26
Horace A. Castaldini 23Aug.27
George J. Ross 13Nov.26
Frederick J. Knowler [e] 18Jan.24
Edwin W. T. Crouch [e] 18Nov.25
Arthur H. Simmonds [e] 27Sept.27
Robert B. Harnden 16Jan.27
John E. L. Drabble 26Mar.27
Charles F. Sealy 27Sept.27
John S. Dick 1Nov.26
James E. Preston 17Oct.26

Stores Branch.
Sqdn. Ldr.
Thomas Bell, *M.M.* 18Nov.25
Flight Lieuts.
James L. Denman 5May23
Hugh Jones 14Jan.27
Ernest W. Lawrence 7Dec.26
Flg. Offs.
Arthur McC. Goddard 14Jan.27
William J. Cleasby 18Oct.24
Leonard T. Sanderson, *D.S.M.* 17Oct.25
George J. Maygothling 15Oct.27
Francis R. Lines 2Dec.27
Thomas I. Iliff 2Dec.27

Accountant Branch.
Sqdn. Ldr.
Herbert F. Fuller 20Apr.27
Flight Lieuts.
Herbert W. Capener 21Oct.26
Alfred W. Gray 27Sept.27
Flg. Off.
John J. T. Rose 1Oct.26

Medical Branch.
Sqdn. Ldr.
Herbert McW. Daniel, *M.D.* 15Oct.27
Flight Lieuts.
Reginald H. Stanbridge 15Oct.27
Albert F. Cook 18Nov.27

Chaplains Branch.
Rev. Denis F. Blackburn 17Oct.26
Rev. William P. Hughes 23Nov.26

Education Officer.
Grade III.
John A. Spencer, *Capt., M.I.Mech.E.* 21Sept.26

AIRCRAFT DEPÔT INDIA.
Karachi.

Wing Cdr
Reginald J. Bone, *C.B.E., D.S.O.* . 16Sept.26
Sqdn. Ldrs.
George C. Bailey, *D.S.O., p.s.a.* [e]. 23July26
William H. L. O'Neill, *M.C., p.s.* .. 20Sept.27
Flight Lieuts.
Kenneth L. Boswell [e] 11Apr.24
William M. M. Hurley 18Nov.25
William F. Dry 19Mar.26
Harold E. Falkner [e] 26Mar.27
John B. H. Rogers 13Nov.27
Flg. Offs.
Percy Coyle 11Dec.26
Frederick Simpson [e] 30Aug.27
William Badley [e] 27Nov.24
Arthur H. Berry, *D.S.M.* [e] 23Nov.23
Michael S. Keogh, *A.M.* 2Oct.25
Edward F. Thorpe (e) 7Dec.26
Gordon H. Bennett 27Mar.27
Charles N. A. B. Mumby 30Nov.26
Hector S. Martin 19Oct.27
William T. Holmes 1Oct.27
Herbert G. Wisher 20Sept.27

Stores Branch.
Sqdn. Ldr.
Percy M. Brambleby 25Dec.26

Aircraft Depôt, India—contd.
Flight Lieuts.
Lamont Smith 13Apr.26
Raymond H. Latham 13Apr.26
Flg. Offs.
John H. P Clarke 27Apr.26
Sidney R. L. Poole 20June27
Charles W. Gore 27Nov.24
Roper G. A. Vallance 27Nov.24
Fred B. Ludlow, *O.B.E., M.C.* 21Sept.26
Geoffrey L. Worthington 11Dec.26

Medical Branch.
Sqdn. Ldr.
Robert S. Overton 20Aug.27
Flight Lieuts.
Brian W. Cross 29Oct.27
John Magner, *M.B.* 17Nov.27

Education Officer.
Grade III.
William F. Foulston, *Capt., M.A.* .. 28Sept.27

AIRCRAFT DEPÔT, 'IRAQ.
'Iraq Command.
Hinaidi.

Wing Cdr.
Victor O. Rees, *O.B.E.* 22Sept.25
Sqdn. Ldrs.
Percy C. Sherren, *M.C.* 2Dec.27
Eric R. Vaisey [e] 18Nov.25
Gilbert D. Nelson, *D.S.C., A.F.C.*[e] 2Dec.27
Tilden C. Thomson [e] 21Sept.26
Flight Lieuts.
Noel V. Wrigley 9Aug.26
Kenneth B. Lloyd, *A.F.C.* 21Oct.27
Allan L. A. Perry-Keene [a] 2Dec.27
Thomas C. Traill, *D.F.C.* [e] 22Sept.25
V.C. Alan Jerrard [e] 20Sept.27
Findlay W. Sinclair, *D.F.C.* [e] 22Sept.25
Ernest C. Barlow 25Oct.27
Flg. Offs.
Thomas Marchant 21Sept.26
David M. Rees, *M.B.E.* [e] 20Sept.27
Charles Snow [e] 21Sept.26
John W. White [e] 22Sept.25
George Lacey [e] 7Dec.26
William J. Coadwell, *D.S.M.* [e] ... 2Dec.27
Francis Fazey [a] 20Sept.27
Robert R. Bennett [s] 21Sept 26

Stores Branch.
Sqdn. Ldr.
Frank Tedman, *M.B.E.* 7Dec.26
Flight Lieuts.
Richard F. Osborne 2Dec.27
Archibald T. Shaw 7Dec.26
Herbert J. Payne 27Apr.27

Flg. Offs.
John J. Ironmonger 10June27
Alfred J. Walker 16Dec.27
Alfred Amy 20Sept.27
Richard H. Clay 21Sept.26
Edward J. Fishenden 21Aug.27
Wilfrid A. D. Collingwood 2Dec.27

HOME AIRCRAFT DEPÔT.
Inland Area, No. 21 Group.

Postal Add.—Henlow Camp, Beds.
Tel. Add.—Aeronautics, Hitchin.
Tel. No.—Hitchin 137-8-9.
Station.—Henlow, via Hitchin, L.M. & S. and L. & N.E. Rlys.

Group Capt.
Charles R. S. Bradley, *O.B.E.* 21July25
Wing Cdrs.
Douglas A. Oliver, *D.S.O., O.B.E.* ... 1Dec.23
Ivor G. V. Fowler, *A.F.C.* [e] 8Mar.26
George W. Williamson, *O.B.E., M.C.* [e] 23Mar.25
John A. G. De Courcy, *M.C.* [e] 20July25
Sqdn. Ldrs.
Joseph Kemper, *M.B.E.* [e] 14Jan.2
Albert E. Pettingell [e] 2Mar.2
Loudoun J. MacLean, *M.C.* 8Aug.27
Franz Workman, *M.C.* 1Mar.2
Flight Lieuts.
Thomas H. Newton, *D.S.C.* 15Aug.2
Arnold S. Thompson 9May2
William B. Everton [e] 11Mar.2
David Drover [e] 1Oct.2
Maurice H. Butler, *D.F.C.* [e] 14Mar.2
Robert W. Edwards 16Aug.2
John Potter 27Aug.2
Tom O. Clogstoun [e] 6Oct.2
Eric Burton [e] 28Sept.2
Leslie G. Harvey [e] 10Sept.2
John Oliver 12July2
John D. Breakey, *D.F.C.* 2Dec.2
Leslie A. C. Stafford [e] 2Oct.2
John Duncan 21Jan.2
Thomas G. Bird 11May2
Frank Jezzard, *M.B.E.*[s] 11Mar.2
Percy I. V. Rippon 14June2
Flg. Offs.
Bertie S. Brice, *A.F.C.* [e] .... 1Sept.
John Bullock 13July
Ernest Whittlesea, *M.B.E.* [e] 1June
Robert A. Whyte .... 15Sept.
Horace T. Satterford [e] .... 1Sept.
William G. Kentfield [e] .... 24Mar.
John W. Caddy [e] .... 3Oct.
William Morgan [e] ....
Sidney Upton [e] .... 14Jan.
Leonard Butler .... 29Dec.
Charles F. H. Grace 15Oct

A page from the RAF List showing officers at RAF Drigh Road

V/1500 bomber that was flown out to India via Egypt. Bombs were successfully dropped on the Afghan King's palace, leading to a hasty armistice.

As part of Trenchard's policy of moving the RAF out to the Empire, three additional squadrons were moved to India, and the total number of squadrons serving there eventually rose to seven by the 1930s. RAF Headquarters was in Simla, with an aircraft park at Lahore and a major repair depot at RAF Drigh Road at Karachi.

Between 1919 and 1925 the RAF flew operations along the frontier in support of an army that controlled the purse strings and, as a result their aircraft were outdated and decrepit. A report in 1922 by Air Vice Marshal Salmond in which he argued for an independent budget and an increase in size was only slowly adopted.

The Government of India was compelled to rethink their traditional attitude of sending expensive (in terms of both money and lives) army expeditions into hostile territory. Between 1925 and 1939 the RAF continued to fly operations against rebel tribes and, in a curious precursor of the situation in the early twenty-first century, they hunted the radical Muslim preacher 'The Fakir of Ipi', who had whipped up trouble among the tribes and hid in caves along the Afghan border and, though hunted for several years, was never captured.

In 1928, in another operation that foreshadowed later RAF operations, the British delegation in Kabul, Afghanistan, as well as several hundred British and other foreign civilians, were evacuated by air following troubles in that country. Using RAF DH9a bombers at first, and later Hinaidi and Victoria troop-carrying aircraft specially brought over from Iraq, a total of 586 civilians were flown out of Kabul via the Khyber Pass. These included the harem of ex-King Ammanulla!

RAF Squadrons based in India 1915–1939

Squadron	*Period*
No. 1 Squadron	January 1921 to November 1926
No. 3 Squadron	February 1920 to October 1921
No. 5 Squadron	March 1920 to Second World War (48 Squadron renumbered)
No. 11 Squadron	December 1928 to Second World War
No. 20 Squadron	May 1919 to Second World War
No. 27 Squadron	April 1920 to Second World War
No. 28 Squadron	February 1920 to Second World War (114 Squadron renumbered)
No. 31 Squadron	late 1915 to Second World War
No. 39 Squadron	January 1929 to Second World War
No. 48 Squadron	June 1919 to March 1920 (renumbered as 5 Squadron)
No. 60 Squadron	April 1920 to Second World War (97 Squadron renumbered)
No. 97 Squadron	July 1919 to April 1920 (renumbered 60 Squadron)
No. 114 Squadron	September 1917 to April 1920 (renumbered as 28 Squadron)

RAF Drigh Road: station records where there is no ORB

RAF Drigh Road, Karachi, in what is now Pakistan, was the main repair depot for the RAF in India between the wars. As well as carrying out major repairs and servicing the aircraft for the whole India Command, many men passed through it on

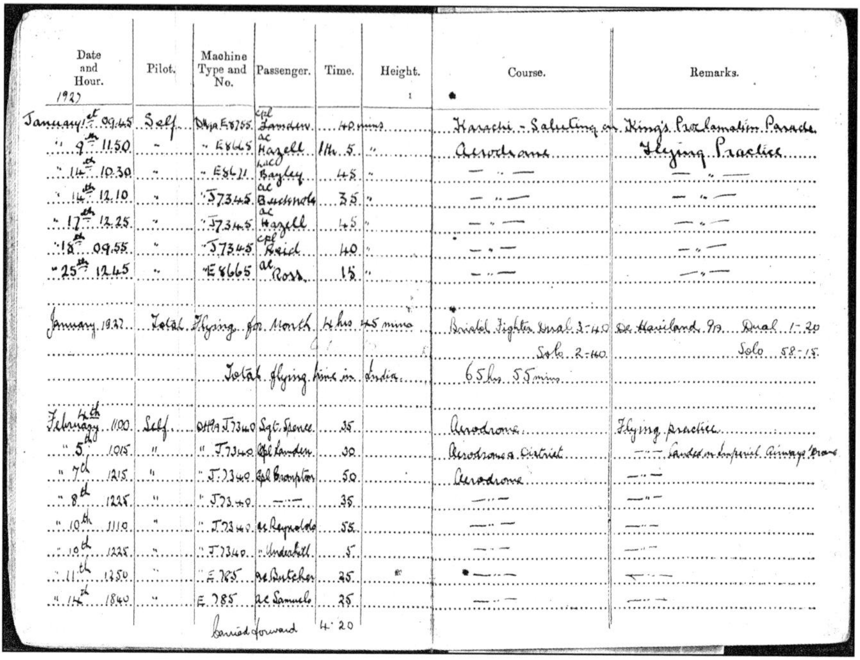

Excerpt from Wing Commander Reggie Bone's log book showing flights made while Station Commander at RAF Drigh Road, Karachi. (Mrs R J Bone via Mrs R Horrell)

their way to stations 'up country' or on their way back to meet the troopship that would take them home. It was also the site for the first international airport in India and was frequently visited by long-distance flights to Australia and the Far East.

There are no station records, in the form of Operation Record Books, for Drigh Road in the inter-war period, and the Group Captain, Pakistan Air Force, who commanded the station in the early 1990s, confirmed to me that there do not appear to be surviving records in Pakistan. A lot of information about activities there can be gained from the 'Royal Air Force in India General Monthly Results Resume and Diary of Operations' which are in AIR 5/1321–37.

Each monthly diary gives the location of the units in the Command, a brief resume of movements and short details of any casualties that have occurred. Each squadron or unit reports on the total hours flown, of cooperation with the Army and on any operations undertaken. Officers are named occasionally but it is rare to find mention of any aircraftsmen by name.

Drigh Road is well served in these diaries so a good picture of the activities can be gained. The diary for February 1927 records 91 hours of flying in the month. Imperial Airways Liner 'City of Delhi' landed on 1 February, carrying the Secretary of State for Air, and left on 3 February. Various aircraft arrived from the squadrons for overhaul and repair and others were sent back. An air survey was carried out for a proposed railway extension, a stocktake was under way and major work was being carried out to establish radio communication with England. On the sporting front, tug of war and running teams were being organized for a forthcoming

The farewell dinner for the radio section, RAF Drigh Road, in 1927. (With thanks to Vic Jones)

RAF radio section at Drigh Road, Karachi, at their ease. AC1 Vic Jones bottom left seated. (With thanks to Vic Jones)

'Assault at Arms', there were poor results in the Mama inter club Tennis Cup, rain damaged the station's Golf Course, the station's hockey teams were beaten in the Hajee Dossal Cup. Six-a-side football was thriving and there had been a successful Inter Section Boxing Tournament, the winners being trained to meet the Sherwood Foresters. Eight aircraft had been fully serviced as well as fifteen engines and a variety of motor transports.

The July 1929 diary records the arrival of the Air Marshal Salmond by flying boat at Karachi harbour and of a long-distance flight from Australia piloted by Captains Kingsford-Smith and Ulm. A Torchlight Tattoo carried out by the Army was accompanied by night-flying demonstrations by an aircraft from the depot.

One of the most famous airmen to serve at Drigh Road was Aircraftsman Shaw (formerly the famous soldier Lawrence of Arabia) and his letters home give an interesting picture of the life of an airman in India.

RAF records for India from First World War to 1939

Of the initial RFC squadrons that served in India there are records of 31 Squadron in AIR 1/1304/204/11/177 (movement of 3 Flight, 31 Squadron to India), some weekly reports of operations on the North West Frontier in 1916 in AIR 1/507/16/3/48, a summary of 31 Squadron work in India in AIR 1/434/15/271/1 and a summary of operations by a detachment of 31 Squadron in Baluchistan in AIR 1/1637/204/92/2. The history of 31 Squadron 1916 to 1917 is in AIR 1/691/21/20/31. There is a history of 114 squadron (1917 to 1936) in AIR 1/695/21/20/114.

There are some Operations Records Books for the RAF squadrons that flew in India in Air 27 series, including 5 Squadron in AIR 27/63 and 28 Squadron in AIR 27/332 and AIR 27/335.

Air Vice Marshal Salmond's highly critical report on the state of the RAF in India from 1922 is in AIR 8/46.

The Air Historical Branch collected and collated much material on operations in India between the wars in what is now TNA's AIR 5 series. Between AIR 5/1321 and AIR 5/1337 are a whole series of papers on RAF actions and summaries of other work undertaken, such as air surveying, air transportation and even air shows. AIR 5/1321 itself contains the supplement to the *London Gazette* covering 'Pink's War', naming those officers and other ranks awarded medals or mentions in despatches, as well as summarizing the operations. Also included in the file is the 'Daily detail of operations by machines of Nos. 1 and 2 Wings Royal Air Force, India during Month of March 1925' which gives a much more detailed account of the war.

Also scattered throughout AIR 5 are various intelligence reports and diaries relating to the North West Frontier on a weekly basis. AIR 23 contains a couple of files on specific local operations during the 1920s and 1930s: AIR 23/683 covers operations in the Afghanistan rebellion 1928–9; AIR 23/688 covers air operations in India 1935 to 1938.

There are various despatches relating to the airlift evacuation of Kabul in AIR 5/725 and AIR 5/726. One of the participants in the airlift, Air Chief Marshal Sir Ronald Ivelaw-Chapman, at the time a Flight Lieutenant, wrote *Wings over Kabul* with Anne Baker (Kimber, 1975) which described his flights to and fro, including a forced landing and being held captive by tribesmen for eighteen days.

28 Squadron lost everything relating to their early service in India during the

retreat from Burma in 1942 but, fortunately for historians, Air Mechanic John Ross (later Flying Officer J Ross) recorded many of his experiences of the early days in his diary, and this has been published in book form, along with as much information as 28 Squadron Association could recover from its members. *The RFC to the RAF, India 1919* was published by Regency Press in 1987 and, as well as John Ross's memoirs. it contains probably the only list of officers and other ranks in existence.

Records of many individual stations abroad have not survived, though those that have are in AIR 29. It is usually possible to find some information about a station by checking for records compiled by the wing, group or command that the unit was part of. These details can be found by checking in the RAF List.

Other records of the RAF in India

A few records relating to the RAF in India survive in the India Office archives held at the British Library.

Royal Indian Air Force and the Royal Air Force in India

Air Force List Dec. 1918–Oct. 1948: L/MIL/17/10/1–300.

Royal Indian Air Force

List 1941–1946: L/MIL/17/10/301–9.
Pay accounts of officers in the UK 1918–1942: L/AG/20/47/1–10. 1943–1947: L/AG/20/35/1–2.

Royal Air Force

Nominal index of personnel serving in India 1919–1939: L/AG/26/12.
Pay accounts of officers from India serving in the UK 1943–1945: L/AG/20/36/1–6.
Unrecovered balances of pay 1943–1950: L/AG/20/37/1–3.

Pink's War: Waziristan 1925 – the RAF's first unsupported war

There are various problems associated with researching the RAF in India between the wars, mainly caused by incomplete or missing records. I've chosen 'Pink's War' (named after the commanding officer, Wing Commander Richard Pink) as an example because it was the RAF's first opportunity to exercise air power on its own against troublesome tribesmen on the North West Frontier

Four groups of tribesmen of the Mahsud tribe on the Afghanistan–Indian border had been making trouble for some years, raiding, cattle rustling, stealing guns, sniping at the army and abducting Hindus. The local Resident asked the Government to sanction action by the RAF.

On 12 February 1925 RAF Headquarters, India, approved a plan of operations and appointed Richard Pink to command. It was decided to employ three squadrons for operations with a headquarters of five officers and twenty airmen to be based at RAF Tank. No. 5 (Army Co-operation) Squadron moved to Tank with ten two-seater Bristol Fighters and 27 (Bomber) and 60 (Bomber) Squadrons were sent to Miramshah with eight DH9a aircraft apiece. The total force comprised 26 aeroplanes, 47 officers and 214 airmen.

The RAF's targets were scattered over an area of 50 to 60 square miles, comprising good-sized villages, scattered huts and cave dwellings of the various tribes, as well as their cattle and sheep. Tactics adopted included bombing by night and day, usually by flights of three aircraft but sometimes involving several flights attacking the targets in succession. Regular and irregular patrols were also flown to attack targets of opportunity and increase the tribes' sense of discomfort and uncertainty. Sometimes attacks were stopped at 3 p.m. to give the impression that they had ceased for the day, but then heavy attacks were launched just before dusk. Bombers cut their engines as they approached their target so as to reduce warning noise to a minimum before dropping their bombs.

Raids were made against villages, cave dwellings, hay stores and herds of cattle. Some delayed action bombs were dropped which continued to explode throughout the night. By the end of March it was becoming obvious that the campaign might be a protracted one and a night-flying Bristol Fighter of 31 Squadron was flown up to Tank. On the night of 4 April it carried out its first attack, which was followed up by others from two more Bristol Fighters brought up to join in. After fifty-four days of constant operations the final group of hostiles surrendered and the first war carried out solely by the RAF had resulted in a signal success.

Researching Pink's War

There are difficulties in researching Pink's operations. There are Operations Record Books for 5, 27 and 31 Squadrons (AIR 27/63, AIR 27/258 and AIR 27/326 respectively) but no ORB for 60 Squadron before 1936.

The ORB for 27 Squadron reads 'No records are available for the period 25.1.24 to 3.6.27'. There are no ORBs or other station records for RAF Tank or RAF Miramshah.

The ORB for 27 Squadron is the most detailed, but even here there are precious few mentions of individuals. Most operations are described briefly, i.e. '22.3.1925 1515 *Bomb raid* 2 machines. Objective Dre-Algad. 1 20lb, 2 230lb & 4 112lb bombs dropped.' The only mention of individuals comes on 21 March 1925 when 'On diving low to attack a gathering of natives machine crashed with bombs on at TORRA TUKKA. Pilot Flying Officer Dashwood & Passenger Flying Officer Hayter-Haymes were both killed. Machine burst into flames and destroyed.'

The ORB for 5 Squadron (which shows every sign of having been written long after the event) records

> After the phase of intensive bombing, which had imposed a considerable strain on both personnel, machines and engines, a system of air blockade was next used, pilots usually working in pairs and the bombing being distributed at uncertain times of the day over the whole area, in order to keep the tribesmen in a continual state of apprehension . . . Gradually the intolerable strain of living in caves, its discomfort, the dislocation of normal life and the inability to graze their stock began to tell on the tribesmen.

A useful source for additional information is the *London Gazette*. A supplement to the *Gazette*, with a description of operations and a list of officers and men awarded medals for gallant and distinguished service and also officers and men 'brought to notice for distinguished service' during the campaign, was published on 17

November 1925. It gives a lot of details about the background to the various disputes that led to the war, the tactics used and some of the major raids carried out. It enlarges slightly on the deaths of Dashwood and Hayter-Haymes and gives details of how their bodies were returned by the tribesmen. It notes that their machine was probably shot down.

Gazettes relevant to specific campaigns can be found by searching the *Gazette* website at http://www.gazettes-online.co.uk/generalArchive.asp. Hard copies of the *Gazette* are in TNA's ZJ series, and you'll often find one in the AIR papers which cover the campaign; the one for Operations in Waziristan (1925) is in AIR 5/1321.

The RAF List will give you the names of all the squadron officers during the campaign and using these names you can check to see whether the RAF Museum or Imperial War Museum hold any donated papers from individuals that might mention your relative.

Other National Archive files which contain information, usually of a high-level nature and not referring to individuals, are:

AIR 5/1330: India: monthly summary of operations and work: Vol II

AIR 5/1321: India and Far East: operations, Chapters 1 to 15

AIR 5/298 Pts.I: Air operations in India, 1922 to 1928

AIR 8/83: The use of air power on the North West Frontier of India: memorandum (1926)

AIR 8/46: India, 1921–1930: reports and papers including report by Air Vice Marshal Salmond on the RAF in India etc.

AIR 30/64: Approval of the India General Service Medal (1908) with Clasp 'Waziristan' for operations in Waziristan, March–May 1925

Wing Commander Reggie Bone in DH9A number E951, probably at RAF Drigh Road. (Mrs R J Bone via Mrs R Horrell)

Other RAF postings abroad

Though Iraq and India were the RAF's principal foreign postings there were others, as well as several long-distance experimental flights made to South Africa and the Far East. 8 Squadron served in Aden for many years and other squadrons were posted to Turkey during the Chanak crisis of 1922 and to China in 1927 as part of the Shanghai Defence Force.

Records for the Chanak crisis, other than those of the squadrons involved, are in AIR 5/250 (Air assistance required for holding the Dardanelles 1921–22), AIR 5/849 and AIR 5/850 (Peace evacuation of the RAF from Turkey, 1923).

Among the records for the Shanghai Defence Force are AIR 5/862: General policy and duties of Royal Air Force Shanghai Defence Force, AIR 5/863: Internal security of RAF Shanghai Defence Force and copies of propaganda distributed by Communists, and AIR 5/866: Daily Intelligence reports of Headquarters RAF Shanghai Defence Force: Nos. 1–70.

There are numerous records of operations in Aden but operations there have been neatly summed up in some monographs by serving officers in the AIR 69 series (compiled for the Staff College at Andover). These include AIR 69/68, *The Work of the Royal Air Force in Aden*, by Squadron Leader RA Cochrane, AIR 69/88, *Small Wars: Aden*, by Squadron Leader J L Vachell, and AIR 69/143: Précis of a lecture 'Air Operations in the Aden Protectorate' by Squadron Leader A H Montgomery.

The RAF blazed the trail for future civil aviation in the Middle East and to India and flew experimental flights to South Africa and Australia. Searches of TNA database under 'long distance', 'experimental' and 'civil aviation', as well as by the destination of the flight, should produce useful results.

The Fleet Air Arm

The amalgamation of the RFC and RNAS into the RAF in 1918 caused one particular problem. What was to be done with the units that were committed to flying off the new-fangled aircraft carriers? It was easy enough to give the RAF bases on land, but the carriers had to be sailed by sailors and to operate as part of a fleet of Royal Navy vessels. Their Lordships at the Admiralty, as the Senior Service, also seem to have resented this new upstart service and began, as soon as the war was over, to agitate for more control over naval aviation. The struggle over this problem was finally resolved in 1938 when the carrier-borne aircraft were given back to the Navy, but for the 1920s and most of the 1930s a sort of unhappy compromise (at least at the top) was the norm.

From 1923 onwards, when the Fleet Air Arm (FAA) of the RAF was officially announced, each aircraft carrier acted as a sort of mobile RAF station, with the aircraft commanded by an RAF officer with his administrative and technical officers also from the RAF. The pilots who served under him were a mixture of RAF officers and RN officers, with the majority (in theory) from the Navy. Naval officers were trained to fly by the RAF and spent some time with the FAA before returning to their normal naval duties.

One thing that the family historian will have to think about is whether their relative served on aircraft carriers as an RAF or RN officer. Fortunately the inter-war RAF and Royal Navy Lists do make it clear which was which. The Royal Navy List for December 1926 shows five FAA Flights aboard HMS *Furious*, under the overall command of Wing Commander H R Busteed, who had five other RAF officers under him as his staff. The six Flights aboard *Furious* had twenty-nine naval officers (including four Royal Marines) as well as thirteen RAF officers, which included all six Flight Commanders. As befits officers of the Senior Service, all the naval officers are listed above the RAF officers in their flights. It is up to the RAF List for the same period to tell us anything about what each Flight was intended to do: 405 Flight was the Fleet Fighter Flight, flying Fairey Flycatchers; 420 and 421 Flights were for Fleet Spotting, and flew Blackburns and Avro Bisons; 443 was a Fleet Reconnaissance Flight equipped with Fairey IIIDs and 461 and 462 were Fleet Torpedo Flights flying Blackburn Dart. In the RAF List the RAF officers are listed above the naval officers!

By the mid-1930s Fleet Air Arm units were redesignated as squadrons with their numbers starting with 8. No. 801 Fleet Fighter Squadron was on HMS *Courageous*, for example, and 823 Fleet Spotter Reconnaissance Squadron was on HMS *Glorious*. Eventually, after twenty years of inter-service disputes, Sir Thomas Inskip, Minister for the Coordination of Defence, produced a paper returning carrier-borne aircraft to Admiralty control, but leaving land-based reconnaissance and attack aircraft under the control of RAF's Coastal Command.

The service record, if you receive it from the RAF or RN, will list the various flights and ships to which your relative was attached. No records for the FAA Flights in the 1920s seem to have survived, but ships' logs for the various aircraft carriers are in TNA's ADM 53 series and can be searched for by name. From 1933 onwards there are ORBs for 701, 720, 801, 802, 803, 805, 806, 812, 813, 815, 816, 819, 821, 823, 824, 825, 826, 827 and 841 Squadrons.

Though these are unlikely to mention individuals by name, there are scores of files on the Fleet Air Arm in both the AIR and ADM series at TNA. AIR 2/490 deals with the reorganization of the Arm on a squadron basis; AIR 8/223 contains Sir Thomas Inskip's report and papers relating to his decision to give the carrier-based aircraft back to the Admiralty; AIR 9/2 is a Directorate of RAF operations and Intelligence paper on the Fleet Air Arm and naval cooperation; AIR 19/96 to AIR 19/105 deal with relations between the RAF and the Navy throughout the 1920s; ADM 116/3417 to ADM 116/3419 are papers of Admiral Lord Keyes on the original organization of the Fleet Air Arm; ADM 116/3430 to ADM 116/3432 deal with the early development of aviation with the fleet; ADM 116/3477 to ADM 116/3479 deal with naval control of the Fleet Air Arm; ADM 116/3001 to ADM 116/3009 deal with various aspects of the reorganization in the early 1930s; ADM 116/3724 to ADM 116/3729 deal with the various practical problems over the return of the Air Arm to Admiralty control.

If your relative was a naval officer who flew with the Fleet Air Arm, their records are yet to be released so you will need to approach the Royal Naval Record Office at NPP Accounts, 1, AFPAA, Centurion Building, Grange Road, Gosport, Hampshire, PO13 9XA for a copy of their service record. If the veteran and spouse are not surviving you will need to either be their next of kin, or have their next of kin's written authority to have information released to you. The RN Record Office will provide you with a Next of Kin form to complete which will explain who the next of kin are (all children count equally for example). It may occasionally be necessary to provide proof of identity. There is a search fee (2007) of £30.

RAF expansion and new commands in the 1930s

On the formation of the RAF in 1918 the organization was based on the original RFC structure of squadrons grouped into wings, wings collected into brigades (later known as groups) and groups organised into areas. In the early 1920s these were geographically based, but faced with a perceived threat from Germany and her rapidly expanding Luftwaffe, the RAF began its own expansion in 1935. As part of this reorganization and expansion the areas became commands based on the role the aircraft were to play.

Bomber Command, Fighter Command, Coastal Command and Training Command were formed on 14 July 1936. On 1 April 1938 Maintenance Command was created, followed by Balloon Command on 1 November that year. Training Command separated into Flying Training Command and Technical Training Command on 27 May. Army Co-operation Command was created on 1 December 1940 (and disbanded on 1 June 1943). Ferry Command was formed (from Atlantic Ferry Organization) on 20 July 1941 and then became Transport Command on 25 March 1943.

As they are the highest levels of RAF organization, papers from the Commands are extremely unlikely to contain information immediately relevant to the family historian but they will be an invaluable source for planning and policy papers for certain operations or reports on incidents.

The Command papers are grouped in TNA series:

Balloon Command	AIR 13
Bomber Command	AIR 14
Coastal Command	AIR 15
Fighter Command	AIR 16
Maintenance Command	AIR 17
Overseas Commands	AIR 23
Air Training Command	AIR 32
Ferry & Transport Command	AIR 38
Army Co-operation Command	AIR 39

The largest groups of papers are in the 'fighting' commands, so that Bomber Command has 4,592 files, Fighter Command has 1,505 and Coastal Command has 986. Maintenance Command has only 118 files, only a few of which actually relate to the Second World War, most coming from the late 1940s and 1950s. Army Co-operation lists 147 files, but many of them relate to the training of paratroops and to paratroop operations in the early part of the war.

AIR 23 is by far the biggest collection of papers, with 8,760 files but this is because it covers overseas commands going back to the early 1920s.

Chapter 6

THE SECOND WORLD WAR: MAIN RAF COMMANDS

The Second World War was the testing and proving time for the Royal Air Force. Its senior officers were men who had fought and flown during First World War (Hugh Dowding, who commanded Fighter Command during the Battle of Britain, we have already met commanding a wing during the Battle of the Somme). Many middle ranking officers had also served in the RFC or RNAS – Reggie Bone, who had seen distinguished service during First World War, in Russia in 1919 and commanded RAF Drigh Road in the 1920s, came out of retirement to serve as a Group Captain commanding RAF Pembroke Dock. The young pilots and ground staff were trained, highly motivated and ready to go. On 3 September 1939 the RAF consisted of 11,529 officers and 162,439 other ranks, supported by 234 Women's Auxiliary Air Force (WAAF) officers and 7,460 WAAF other ranks. On 1 September 1945 there were 100,107 RAF officers, 840,760 other ranks, 5,638 WAAF officers and 135,891 WAAF other ranks, though tens of thousands more had seen service and been killed or discharged.

In the first few months of the war an RAF force was sent to France to support the army and there were occasional reconnaissance flights by German aircraft to deal with. Bomber Command began daylight attacks on the German fleet and naval bases – these raids were not a success and suffered heavy casualties, causing Bomber Command to switch to night bombing. Their failure did point out something previously considered impossible – given warning, fighters could intercept and badly maul bombers before they reached their targets. Fighter Command took note and began to reconsider their tactics – just in time.

Fighter Command

Fighter Command was formed in July 1936 under Air Marshal Sir Hugh Dowding, with its headquarters at Bentley Priory, Stanmore, Middlesex. Its main role was the air defence of Britain and Dowding built a superb organization, with Fighter Groups defending specific regions, and sector control operation rooms controlling groups of fighters by voice radio. Close links with the Observer Corps and the radar

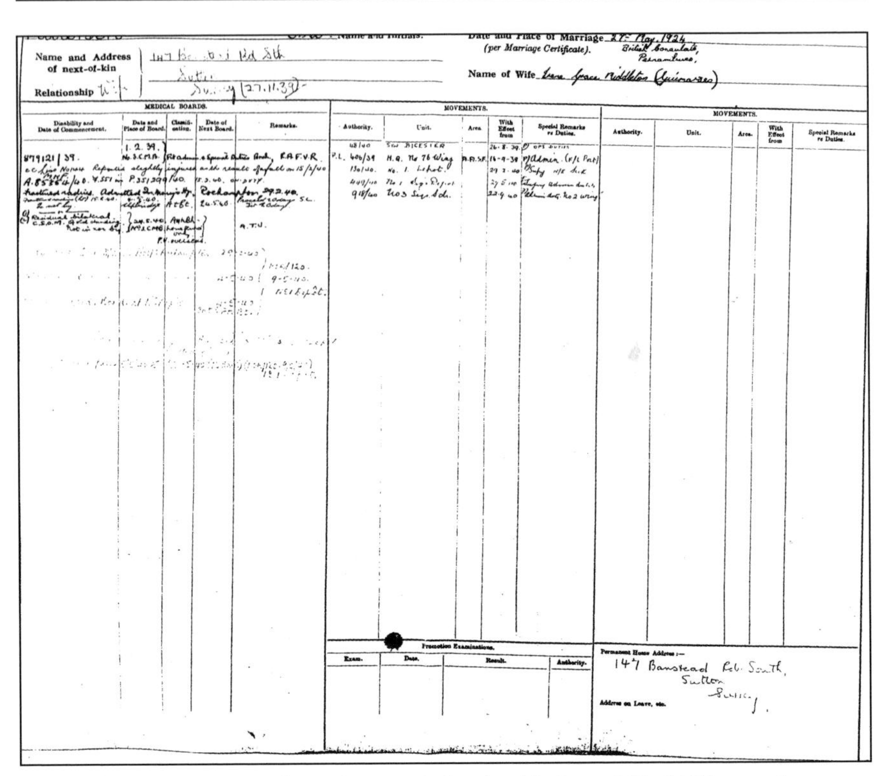

Name and Address of next-of-kin: 147 Banstead Rd Sth, Sutton, Surrey (27.11.39)

Relationship: Wife

Date and Place of Marriage (per Marriage Certificate): 27 May 1924, British Consulate, Pernambuco

Name of Wife: Irene Grace Middleton (Guimaraes)

MEDICAL BOARDS.

Disability and Date of Commencement.	Date and Place of Board.	Classification.	Date of Next Board.	Remarks.

MOVEMENTS.

Authority.	Unit.	Area.	With Effect from	Special Remarks re Duties.
43/40	Stn Bicester		26.8.39	
P.L. 406/39	H.Q. No 76 Wing	A.A.S.F.	14-9-39	Admin.
130/40			29.2.40	
449/40			27.5.40	
918/40			22.9.40	

MOVEMENTS.

Authority.	Unit.	Area.	With Effect from	Special Remarks re Duties.

Promotion Examinations.

Exam.	Date.	Result.	Authority.

Permanent Home Address:— 147 Banstead Rd. South, Sutton, Surrey.

Address on Leave, etc.

Excerpt from J W Patterson's service record showing his Second World War service

stations, Balloon Command and anti-aircraft defences created a sophisticated defence system. Sector control rooms were able rapidly to move scarce aircraft to meet the threat of Luftwaffe attacks and counter the enemy's numerical superiority.

During the Battle of Britain it was the resilience and flexibility of Fighter Command's organization and communications and the hard work and ability of the men and women at ground level who maintained the aircraft and aerodromes and kept the communications open that kept 'The Few' in the air. Their contribution to the battle should not be overlooked.

Fighter Command files (including those from 1943/4) are in the AIR 16 series. Many of the earliest records, from 1936 to 1939, deal with the setting up of the command, its structure, proposed tactics and methods of fighter control, as well as the use of the new radar and liaison with the Observer Corps and Balloon Command.

Battle of Britain: records for a fighter squadron airfield – RAF Hornchurch

With the fall of France in 1940 Germany was free to turn her attention exclusively to Britain. With the BEF rescued, but disorganized and ill-equipped, the short hop

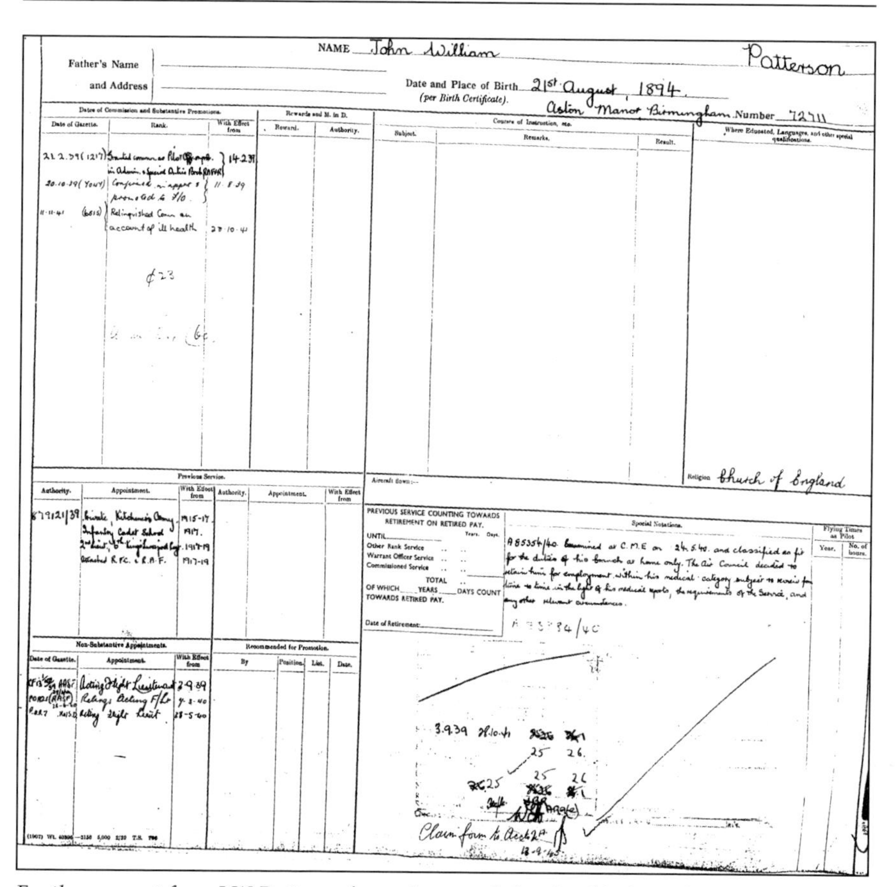

NAME John William Patterson

Father's Name and Address

Date and Place of Birth 21st August 1894. (per Birth Certificate). Aston Manor Birmingham

Number 72711

Dates of Commission and Substantive Promotions.

Date of Gazette.	Rank.	With Effect from
21.2.39 (1217)	Granted commn as Pilot Officer in Admin. & Special Duties Branch RAFVR	14.2.39
20.10.39 (7044)	Confirmed in appt & promoted to F/O.	11.8.39
11.11.41 (6515)	Relinquished Comm on account of ill health	27.10.41

Rewards and M. in D.

Reward.	Authority.

Courses of Instruction, etc.

Subject.	Remarks.	Result.

Where Educated, Languages, and other special qualifications.

Religion Church of England

Aircraft down:—

Previous Service.

Authority.	Appointment.	With Effect from
879121/39	Private, Kitchener's Army	1915-17
	Infantry Cadet School	1917.
	2nd Lieut. 6th King's Liverpool Regt.	1917-19
	Attached R.F.C. & R.A.F.	1917-19

Non-Substantive Appointments.

Date of Gazette.	Appointment.	With Effect from
CF 13/5/39 AASF	Acting Flight Lieutenant	3.9.39
PO RD1 (AASF) 29/4/40	Relinqs Acting F/Lt	4.3.40
26-6-40 R.ER.7	Acting Flight Lieut	28-5-40

Recommended for Promotion.

By	Position.	List.	Date.

PREVIOUS SERVICE COUNTING TOWARDS RETIREMENT ON RETIRED PAY.

UNTIL — Years. Days.
Other Rank Service
Warrant Officer Service
Commissioned Service
TOTAL
OF WHICH — YEARS — DAYS COUNT TOWARDS RETIRED PAY.

Date of Retirement:

Special Notations.

A85354/40 Examined at C.M.E on 24.5.40. and classified as fit for the duties of his branch at home only. The Air Council decided to retain him for employment within his medical category subject to review from time to time in the light of his medical reports, the requirements of the Service, and any other relevant circumstances.

A 85354/40

3.9.39 28.10.41

Claim form to Acct 2A 18-9-41

Flying Times as Pilot — Year. No. of hours.

Further excerpt from J W Patterson's service record showing his Second World War service.

over the Channel did not look impossible. Preparations began along the French and Belgian coasts for the invasion, code-named 'Operation Sealion'. The Royal Navy were still a formidable threat to a seaborne force so it was necessary to gain air superiority over the Channel and South East England to cover the initial landings. Once ashore, few doubted the German Army would soon sweep through the battered remnants of the British Army and the newly formed Home Guard. Hugh Dowding's Fighter Command were the front line of the battle, though it should be remembered that everyone in the country took their small part in the Battle of Britain, even if just 'Digging for victory' to provide extra food, firewatching or making cups of tea for the Civil Defence.

There are plenty of good histories of the Battle of Britain available so I've decided just to look at one RAF station, chosen more or less at random from the ones in the forefront of the fray. RAF Hornchurch was part of 11 Group and home to the Hornchurch sector operations room and staff, and the following squadrons during the battle (sometimes for only a few days) :

65 Squadron June to August 1940

74 Squadron June to August 1940

54 Squadron April/May, June/July and August 1940 (three visits)

41 Squadron July, August and September 1940 (three visits)

92 Squadron June 1940

222 Squadron May and June 1940 and late August onwards

266 Squadron from 14 August 1940

600 Squadron late August and early September 1940

264 Squadron late August 1940

603 Squadron from 27 August 1940 onwards

Though not absolutely complete, the following files relate to either Hornchurch itself, 11 Group, or to the squadrons that served there during the battle:

AIR 2 RECORDS: AIR MINISTRY REGISTERED CORRESPONDENCE

Air 2/5246: Attacks on England from 11 September–31 October 1940: No. 11 Group reportAir 16 Records: Fighter Command Registered Files

Air 16/212: No. 11 Group operational Orders, November 1939 to September 1941

Air 16/330: Reinforcement of No. 11 Group

Air 16/352: No. 11 Group: operations over France, May to June 1940

Air 16/635: No. 11 Group: activities

Air 16/870: No. 11 Group: operations over France, May to June 1940

Air 16/901: No. 11 Group instructions

Air 16/839: No. 11 Group: air combat results chart, May to November 1940

Air 16/635: No. 11 Group: activities

Air 16/856: Squadron Combats, RAF Hornchurch, Essex

Air 16/1032: Honours and Awards, RAF Hornchurch

Air 27 Records: Squadron Operational Record Books

Air 27/592: 65 Squadron, September 1934 to December 1940

Air 27/640: 74 Squadron, September 1935 to December 1940

Air 27/511: 54 Squadron, January 1930 to December 1940

Air 27/424: 41 Squadron, April 1916 to December 1940

Air 27/1558: 266 Squadron, October 1939 to December 1942

Air 27/2059: 600 Squadron, January to December 1940

Air 27/1553: 264 Squadron, October 1939 to December 1945

P/O GLASER. 65 Squadron. 37

Attacked : 1 Me.109.

Total Rounds Fired: 2800.

Result : 1 Me.109 Probable.

Whilst flying on patrol on the 22nd of August, in position 2 of Red Section, we engaged the enemy at 25,000 feet.

I engaged one Me.109 and gave a 4 second burst from 250 yards, whereupon, the e/a half-rolled down ; I followed, and when it had regained level flight I gave another short burst followed by a second and longer one.

The e/a took no evasive action and during the second burst there was a flash of flames from the port side of the fuselage, and smoke poured out ; he then continued on a shallow dive and disappeared into the clouds which were at about 10,000 feet. Just before he entered the cloud I overshot him and did a steep turn above him and observed the e/a to be a mottled grey, with orange wing-tips and the usual black crosses outlined with white. I later found I was over the French Coast.

Glaser P/O.

Combat report from Pilot Officer Glaser of 65 (East India) Squadron based at Hornchurch during the Battle of Britain. (TNA AIR 50/25)

Air 27/1555: 264 Squadron, Appendices, May 1940 to April 1943

Air 27/2079: 603 Squadron, September 1925 to December 1943

Have a care when searching for Squadron Operation Record Books using TNA search engine – ORBs for the war period are usually grouped as a Sub Series record and you'll need to click on the Sub Series header and then 'Browse from here'. This will show you all the records.

AIR 28 RECORDS: OPERATIONS RECORD BOOKS, RAF STATIONS

Air 28/384: Operations Record Book, RAF Hornchurch, October 1915 to December 1941

AIR 50 RECORDS: COMBAT REPORTS, SECOND WORLD WAR

Air 50/21: 54 Squadron, Combat Reports, February 1940 to November 1941

Air 50/25: 65 (East India) Squadron, Combat Reports, May 1940 to April 1945

Air 50/32: 74 Squadron, Combat Reports, November 1939 to November 1945

Air 50/18: 41 Squadron, Combat Reports, October 1939 to May 1945

Air 50/105: 266 (Rhodesia) Squadron, Combat Reports, June 1940 to April 1945

Air 50/164: 600 Squadron, Combat Reports, May 1940 to April 1945

Air 50/104: 264 Squadron, Combat Reports, May 1940 to March 1945

Air 50/167: 603 Squadron, Combat Reports, October 1939 to February 1942

AIR 20 RECORDS: PAPERS ACCUMULATED BY THE AIR HISTORICAL BRANCH

Air 20/2872: No. 11 Group – Reports on air fighting, November–December 1940

There are several Battle of Britain pilots' flying log books in the AIR 4 series.

It is all too easy to think of the Battle of Britain as only being fought by Fighter Command. Correctly it is 'The Few' that are remembered first but, of course, other commands did their bit. Bomber Command flew countless sorties against invasion barges in the Channel ports and attacked German aerodromes when it could; Coastal Command flew reconnaissance missions against German shipping and patrolled the sea lanes. When the shortage of pilots became acute they were loaned to Fighter Command from other commands and the Fleet Air Arm.

There are several websites dedicated to the Battle of Britain.

- http://www.the-battle-of-britain.co.uk contains, among other things, a list of RAF pilots who took part in the battle.
- http://www.raf.mod.uk/bob1940/bobhome.html is the Royal Air Force's own web page devoted to the battle, which includes a Roll of Honour, a day by day high-level diary of events giving statistics of casualties, patrols, barrage balloons flown, lists of raids, movements of squadrons and weather conditions. There are Home Security reports giving brief details of air raid damage and brief biographies of the senior officers in both the RAF and Luftwaffe.
- http://www.battleofbritain.net is the official website of the Battle of Britain Historical Society and gives splendid details of the action day by day, including descriptions of the fighting by participants of both sides. There are features on the aircraft that took part, not just the Spitfire and Hurricane but also the Gloster Gladiator and Boulton Paul Defiant, two older fighters that covered the West Country, as well as German aircraft. There are documents on the development of radar, on the Observer Corps and on Air Intelligence ('the eyes of Fighter Command').

(*3865—1660) Wt. 7403—602 5,000 7/38 T.S. 118

OPERATIONS RECORD BOOK

See instructions for use of this form in K.R. and A.C.I. and notes in R.A.F. Field Service Pocket Book.

R.A.F. Form 540

of (Unit or Formation) NO.65 'EAST INDIA' SQUADRON.

No. of pages used for day 3.

Place.	Date.	Time.	Summary of Events.	References to Appendices.
HORNCHURCH	14.8.40	-	The results of the day's operations were again very satisfactory - our claims being 2 Destroyed and 2 Probables, with no loss to our personnel. One of our aircraft was hit by 6 cannon-shells in the wing and fuselage, but the pilot (PILOT OFFICER L.L. PYMAN) effected a successful forced landing and the machine was only temporarily unserviceable.	
"	15.8.40	-	The Squadron commenced flying at 05.45 hours and were engaged on operations the whole day, but there was nothing of importance to report.	
"	16.8.40	-	Operating from MANSTON the Squadron were again heavily engaged on War Operations resulting in our claiming 4 Destroyed, 2 Probables, 2 Damaged (See Intelligence and Combat reports attached). After this engagement PILOT OFFICER L.L. PYMAN failed to return to base and was reported Missing.	
"	17.8.40	-	Convoy Patrols, and War Operations were carried out during the day but nothing of importance was reported.	
"	18.8.40	-	From 0530 hours to 1830 hours the Squadron were engaged on Fighter Patrols and various other War Operations. At approximately 1300 hours while patrolling CANTERBURY and MANSTON, a lone HEINKEL 111 was attacked and probably destroyed as 4 of our pilots reported firing at it (See Intelligence Report attached). At 1250 hours one Flight were ordered off from ROCHFORD, the results of this operation were not reported, but from this patrol FLYING OFFICER GRUSKA failed to return and was reported missing.	
"	19.8.40	-	The Squadron were operating from ROCHFORD and MANSTON during the day and carried out operational patrols and War Operations, but there was nothing reported. PILOT OFFICER CHAPPEL and SERGEANTS HINE, MITCHELL, and STILLWELL were posted to this Unit from No.5 Operational Training Unit, for Flying Duties.	
"	20.8.40	-	Operating from ROCHFORD the Squadron were engaged in War Operations, resulting in our claiming the following :- 2 ME.109's,(DESTROYED) 1 DO.17 and 1 ME.109 (PROBABLY DESTROYED), 8 other enemy aircraft were attacked but no claims made.	

Excerpt from the Squadron Operations Record Book for 65 (East India) Squadron based at RAF Hornchurch during the Battle of Britain. (TNA AIR 27/592)

- http://www.battle-of-britain.com is a nice general site with some good photographs.
- http://www.deltaweb.co.uk/bbmf/home.html is the official site of the Battle of Britain Memorial Flight, the RAF's commemorative flight, with some excellent information on, and photographs of, the aircraft of the Flight.

Bomber Command

Bomber Command was formed on 14 July 1936 under the command of Marshal of the Royal Air Force Sir Edward Ellington. He laid the foundations of later heavy bomber campaigns by ordering the first four-engined heavy bomber, the Stirling. The command began the war staging daylight air raids on the German fleet's bases and suffered heavy losses from German fighters, forcing it to turn to the much less accurate night bombing.

With the Battle of Britain over the RAF considered how it could carry the war to Germany. Bomber Command was given the task of attacking Germany's industries and cities and massively expanded. Halifax, Stirling and Lancaster bombers carried out night raids on a grand scale. The first 1,000 bomber raid on Cologne took place on 31 May 1942 and dropped 1,455 tons of explosives and incendiaries. On 25 June 1942 1,000 bombers levelled the town of Bremen. Huge raids took place almost nightly. Berlin was struck several times. After the USA entered the war her bombers raided Germany during the day. Immense damage was done to German industry and hundreds of thousands of men and guns were diverted to her defence that could have been used elsewhere. The 'Dambusters Raid' of 16/17 May 1943 did great damage when the Mohne and Eder dams were breached by 617 Squadron and a precision raid on the scientific establishment at Peenemunde on 17/18 August delayed research into the V2 rocket by months. The United States Air Force bombed Germany by day and Bomber Command attacked by night, devastating German cities and industry, right up to the end of the war. Over 55,000 Bomber Command crewmen were killed in action.

Bomber Command files are in TNA's AIR 14 series and contain planning files on particular operations, training schemes, tactics, technical developments, selection of personnel and intelligence on targets.

Bomber Command crew log books will often refer to specific operations by their codeword. Operation Sunrise, on 24 July 1941 was, for example, a major bombing raid on the enemy battleships at Brest by a force of over a hundred Wellingtons, Hampdens and Boeing Fortresses. The planning file for this operation is in AIR 14/637. The controversial raid on Dresden in February 1945 was Operation Thunderclap and there are planning files in AIR 20/4826, Operation 'Thunderclap': bombing attacks to break German morale: aerial photographs, and AIR 20/4831.

Operation Millennium: the thousand bomber raid in Cologne, 30/31 May 1942

On 22 February 1942 Air Marshal Arthur Harris took command of Bomber Command. He was a forceful personality who had seen air power at first hand during his period in Iraq in the 1920s and he was convinced that bombing could bring Germany to her knees. It had already been decided, during 1941, and long before Harris took charge, that strategic bombing of Germany on a huge scale was to be the aim of Bomber Command. Harris accepted the strategy and worked hard to see that it was carried out.

Bomber Command had a normal operating strength of about 300 aircraft at any one time, in thirty-eight operational squadrons. On the night of 28/29 May 1942 they successfully bombed the town of Lubeck, doing great damage and then four consecutive nights raids on Rostock, doing huge damage with incendiary bombs. Harris, knowing the value of publicity, resolved that a headline-making raid by a force of a thousand bombers on a major German city would put Bomber Command on the map and deal a heavy blow to enemy morale. By carefully hoarding his own command's planes he could put together a strike force of some 600 aircraft and he set about borrowing aircraft and crews from other commands. An initial promise of bombers from Coastal Command was vetoed by the Admiralty, who had gained operational control of the command during 1941, but 367 aircraft were provided by training groups, and others from conversion flights. 1,046 aircraft took off to attack Cologne on the night of 30/31 May 1942.

Flying in a bomber stream of three waves, led by Wellington and Stirling aircraft fitted with new radio guidance systems that helped navigators locate their position, the first incendiary bombs rained down on the city shortly after midnight. Within a quarter of an hour the old town at the heart of the city was ablaze, providing a marker for the aircraft that followed, who peeled away to bomb the industrial areas on the outskirts. The German civil defence was slow to respond and over 600 acres of the city were burnt down, including 13,000 homes destroyed and 6,000 damaged. Less than 500 people were killed, but nearly 5,000 were injured and 45,000 made homeless. For a week the city lay paralysed, but within six months it was working once again.

Bomber Command lost forty-one aircraft, mainly to German night fighters, but this was considered a reasonable level of casualties. Churchill wrote to Harris: 'This proof of the growing power of the British bomber force is also the herald of what Germany will receive, city by city, from now on.'

Finding out about the raid

Almost the whole of Bomber Command as it was in May 1942 took part in the raid and there is an Order of Battle for the command at that time in Appendix 6.

AIR 14/620 contains the 1 Group Operational tactical reports and summaries for 1942 and notes that, after good weather early in May 1942, in which five operations were carried out, unsuitable conditions for most of the rest of the month severely restricted the group. In fact this bad weather, restricting flying and operational losses, may have been one factor in allowing Harris to marshal the force of over 1,000 bombers for the raid. The Group Tactical Summary goes on to say:

> The outstanding operation of the month was the raid on Koln on the night of 30/31 May, when 1042 aircraft attacked this important industrial and residential city. Weather conditions were very good. The Group, reinforced by instructors and selected personnel from 91 Group, despatched 181 aircraft, made up of 154 aircraft from the Group and 27 from the O.T.U's. Of the 184 aircraft detailed, only 3 were cancelled, a fact which reflects great credit on the squadron ground personnel. The operation was a complete success right from the take off. The ground organisation was excellent, timing was particularly good and signals procedure worked smoothly. The Armament Sections, who had a formidable array of bombs to deal with, deserve special commendation. 157 sorties including 24 O.T.U. sorties attacked the objective with excellent results.

It is worth noting that group appreciated the splendid work done by the ground crew in preparing the aircraft for the battle. Only a very small number of RAF personnel actually flew in combat, and it required something like 140 people on the ground to get the machine and crew airworthy and to keep them that way.

The Operational Order for the raid for 103 Squadron is in AIR 27/819. It is clearly stated 'INTENTION: To cause maximum damage at point of aim'. 103 Squadron were based at RAF Elsham Wolds and the Operations Record Book for the base (AIR 28/255) records:

> Twenty nine aircraft (nineteen of 103 Squadron and 10 from 22 O.T.U.) took part in the great raid on Cologne. Two failed to return (F/Sgt Onions, 103 Squadron and F/O Hannan, 22 O.T.U. In addition F/Lt Saxelby's second pilot, Sgt Roberts, was killed and his navigator and rear gunner injured in an encounter with an enemy night fighter on the return journey. Another

> aircraft, having landed at Kirmington on its return, took off again to complete its journey but crashed; the captain, Sgt Flowers, and two other members of the crew, were killed. Twenty six of the aircraft attacked the primary target; one abandoned its mission due to an electrical fault.

The 103 Squadron report on the raid is in AIR 237/814 which is the Operations Record Book (ORB) covering the whole of 1942. As well as listing the aircraft and crews that took part it gives a brief summary of what the crews reported on their return, or gave what little information was available about their fate if they did not return. After stating that 19 aircraft were 'detailed to cause maximum damage to COLOGNE, point of aim "A"' the ORB confirms the details given in the station log and adds some more. 'Flt Lt Saxelby bombed the target and on departure saw many fires burning. He force landed at Honington. His aircraft had been badly damaged after being attacked by an enemy night fighter, during which action his second pilot was killed.'

There are no combat reports extant for the aircraft of 103 Squadron, but a report from a Wellington of another 1 Group Squadron, number 150 Squadron, in AIR 50/220 gives a good picture of the nature of air combat during the raid. The action took place about 10 miles south of Eindhoven at 01.40 hours at a height of 17,000 feet in good visibility, no cloud, with a full moon. An enemy ME 110 was sighted slightly below, approaching on a reciprocal course (i.e. coming straight at the Wellington) at a distance of 200 yards:

> There was no searchlight or flak activity at the time and IFF had not been used. Although our aircraft took normal evasive action, altering course 30 degrees to port and starboard alternatively, the ME 110 turned and closed in to attack from the starboard quarter at same height. Our rear gunner Sergeant Dent, fired one seven seconds burst at 400 yards range and enemy aircraft dived and was lost sight of temporarily.
>
> Five minutes later the ME 100 appeared once again on the starboard quarter, and attacked our aircraft closing in from 300 yards range, firing three short bursts, both machine guns and cannon. Our rear gunner replied and claims in which he is supported by the Second Pilot, to have secured 'belly' hits at 80 yards range. Enemy aircraft dived from view and was not seen again. At this moment our aircraft went into an evasive dive from 17,000 to 12,000 feet so the enemy aircraft was not seen to crash, but it was probably damaged or destroyed.

AIR 14/1378 contains aerial photographs of the bomb and fire damage in Cologne, as well as maps of the city.

AIR 50/213 contains an account of a combat between Wellington 111 A/C 'B' of 115 Squadron and single-engined fighter during the Cologne raid, 00.40 hours, 14,000 feet, 23 April 1942. Crew: Captain Pilot Officer Patterson; Second Pilot Sgt Fry; Navigator Sgt Robson; Wireless Operator Sgt Williams; Front Gun Sgt Skinner; Rear Gunner Sgt Bourne.

> Wellington Mark III 'B' of 115 Squadron was flying at Air Speed 185 mph on a course of 300 degrees magnetic when a single engined fighter probably a FW 190 approached from the green quarter slightly below at a distance of 250/300 yards. The enemy aircraft opened fire at a distance of 250 yards with machine guns in each wing. Wellington then turned to green and our rear gunner

Coastal Command ground crew moving a Sunderland flying boat at Pembroke Dock, 1940. (Mrs R J Bone via Mrs R Horrell)

> replied from about 250 yards with 2 short bursts of around 40 rounds from each gun. Enemy aircraft passed below to red and disappeared. There was 7/10 to 8/10 cloud at the time of the encounter, and Wellington was not coned at the time of the attack, although searchlights were operating on either side. Wellington was later held at 0110 hours by searchlight for about 1 minute.
>
> Flak was observed at some distance from either beam and immediately below. No IFF.

Coastal Command

Coastal Command Registered files relating to organization, planning equipment and operations are in AIR 15 series at TNA, and some go back into the 1930s when the organization was still known as Coastal Area. The command was finally wound up in 1969 and became part of Strike Command, so the files in AIR 15 continue well into the 1960s.

The command was responsible for land-based long-range reconnaissance of the sea lanes, as well as for land-based bombers and torpedo bombers of the RAF flying in support of the Royal Navy. Their flying boats (originally mainly Short Sunderlands and later Catalinas) escorted convoys and made independent reconnaissance and anti-submarine patrols. In the early part of the war the command even used De Havilland Moth trainer aircraft for coastal patrols, such was the shortage of mainline aircraft. As the Battle of the Atlantic raged, Coastal Command used long-range bombers to carry the anti-submarine war into the very centre of the Atlantic. Their Beaufighters and other aircraft ranged far and wide, attacking enemy shipping with cannon, rockets and torpedoes.

As with other RAF commands, squadron records are in AIR 27 and station records in AIR 28 as usual.

Because of the close liaison with the Admiralty (who actually took over operational control of Coastal Command in 1941), it is always worth searching the ADM files for references. Most of the Admiralty files deal with policy matters, but some operational reports, which may contain names of individuals, do exist.

ADM 1/11061: Transfer of Coastal Command Operational Control from Air Ministry to Admiralty: minutes of meeting and report of Committee on Coastal Command, 1941

A group of naval officers who made up the liaison section at Pembroke Dock in 1940, with Group Captain Reggie Bone. (Mrs R J Bone via Mrs R Horrell)

ROYAL AIR FORCE.

AIRMAN'S RECORD SHEET (Active Service—Overseas)

Official No. S K 4 Name WHEELER, Charles Rank W.O
(In full in block capitals, surname first). (Or acting appointment).

R.A.F. Trade Fitter General Special Qualifications
(e.g., Gas Instructor, Fire Fighter, Boxing Instructor).

Date of Birth 21.6.1886 Religion C of E Occupation in Civil Life Fitter

Last Enlisted 5.8.1914 Current Engagement 24 years

If Reservist, which Class ("E," "F," S.R.) Whether Married, Single, or Widower M

Name and Address of Person (or Persons) to be informed of Casualties:

(In pencil).

Relationship

(In pencil).

Any alteration to above (e.g., Promotions) to be made by crossing out and writing above.

Unit FROM which	Unit TO which	Date of Effect	Authority	Description	Date of Effect
			CP 4/35	Promoted Warrant Officer	27.1.35

Excerpt from the Second World War RAF service record of Warrant Officer Charles Wheeler. (Author's collection)

SECTION 4.—CHARACTER AND TRADE PROFICIENCY.

(To be assessed on every occasion on which an airman is struck off the strength of a unit, e.g., on posting, admission to hospital, death, etc.)

Rank.	Character.	Trade Classification.	Proficiency.			Whether specially recommended, recommended, or not recommended for promotion or reclassification.	Date.	Signature and Rank of Commanding Officer.
			A	B	C			
W.O	V.G.	F.A.E		Ex	Supr		31.12.38	Sgd. H.S.P. WALMSLEY G/C
W.O.	VG	F.A.E	—	Ex	Ex		31.12.39	[illegible] G/C
W.O.	VG.	F.A.E.	—	Ex	Ex	—	31 Dec 40	[illegible] S/Ldr
W.O.	VG.	F.A.E.	-	Ex	Ex		31.12.41	[illegible] Capt
W O	VG	F II E	—	Ex	—		31.12.42	[illegible] S/L
WO	VG	F II E	-	Ex	-	N.R.	5.5.43	[illegible] S/L

PRIV. 22/8 28/8/43 4 169/43 3200 ISSUED 24/8/43

48 HOUR PASS 10-11/4/1943. POR 131/43

29/5 31/5/43 3 108/43 T

P.O.R. 38/43 FROM 23/4/43 TO 26/4/43 (4 DAYS PL) R. WARRANT NOT ISSUED

P.O.R. 64/43 FROM 1/3/43 TO 3/3/43 3 PL R. WARRANT NOT ISSUED

P.O.R. 310/42 FROM 23/12/42 TO 28/12/42 7 PL NOT

P.O.R. 157/42 FROM 15/9/42 TO 18/9/42 (4 DAYS PL) R. WARRANT NOT ISSUED

P.O.R. 138/42 FROM 18/8/42 TO 20/8/42 (3 DAYS PL) R. WARRANT NOT ISSUED

P.O.R. 195/42 FROM 14/7/42 TO 17/7/42 (4 DAYS PRIV) R. WARRANT NOT ISSUED

7 days 21.5.42 27.5.42 POR 138/42 Priv Leave

4 days 22.3.42 25.3.42 POR 80/42 P.L.

3 days 23.1.42 25.1.42 POR 23/42 P.L. RW

7 days 22.12.41 28.12.41 POR 333/41 P.L. RW

7 days 23.6.41 29.6.41 POR 146/41 Priv leave

3 days 18.4.41 20.4.41 POR 95/41 Priv leave

4 days 10.3.41 13.3.41 POR 59/41 Priv Leave

4 days 24.12.40 27.12.40 Priv POR 326/40

7 days 2.11.40 8.11.40 POR 291/40 Priv Leave

7 days 17.8.40 23.8.40 P.O.R. 240/40 Priv Leave.

Annual leave 23:12:39 to 27:12:39 (5 days) no warrant issued

Privilege leave 24.2.40 to 26.2.40 (3 days) no warrant issued

SECTION 5.—DECORATIONS, MENTIONS, SPECIAL COMMENDATIONS BY A.Os.C., ETC.

Authority.	Nature.	Date of Effect.
CF 26/33	Awarded Long Service & Good Conduct Medal	5.8.1932
London Gazette	awarded Mentioned in Despatches Certificate	1.1.42.

NOTE.—This Form is to be forwarded immediately, together with Form 121, to the Commanding Officer of the Unit (if within the Command) to which the airman is posted whilst effective.

It is to be forwarded immediately, together with Form 121, to the B.P.S.O. in all other cases where the airman is struck off strength of his Unit, e.g., killed, died, admitted to hospital, transferred to Home Establishment, posted outside the Command, etc.

Excerpt from the Second World War RAF service record of Warrant Officer Charles Wheeler. (Author's collection)

ADM 1/9535: Coastal Command and air reconnaissance over the sea: organisation, 1937–1939

ADM 1/13695: Air attacks on U-boats at sea: co-operation with Coastal Command, 1943

ADM 219/633: Coastal Command anti-U boat operations 1942–1945

ADM 116/4869: Coastal Command: transfer of operational control to Admiralty, 1940–1943

ADM 199/1111: Coastal Command: narratives and reports, 1945

ADM 199/1691: Coastal Command Liaison Section: Coastal Command policy, training, ferrying, planned flying and maintenance, squadron histories, etc, 1941–1945

ADM 199/1697: Coastal Command Liaison Section: Anti-submarine sweeps, patrols, Air-Sea Rescue, air transport, 1941–1946

ADM 1/17603: Precis of attack on U-boat by Sunderland aircraft, 1945

The RAF in other theatres

Though I have concentrated on the RAF based in the UK, they did, of course, serve all over the world and fought with great gallantry in North Africa, Greece, Italy and the Mediterranean, in India, Ceylon, Burma and the Far East. Some squadrons of Hurricanes went to North Russia. Aircraft patrolled the Atlantic from bases in West Africa, the Caribbean, Iceland and Newfoundland. Training units were established in Canada and South Africa. Group and wing records are again in AIR 25 and AIR 26, and AIR 27, AIR 28 and AIR 29 for squadrons, stations and miscellaneous units respectively. Foreign Command records are in AIR 24.

Chapter 7

OTHER RECORDS FOR THE SECOND WORLD WAR

The RAF Confidential List

In early 1939 the Air Ministry finally woke up to the fact that the Air Force List, available to anyone who wanted one from HMSO at the cost of one shilling, could be used by a potential enemy to identify new squadrons, their bases, their aircraft and their officers. The list was still published, but in two parts. The first continued to list officers, their ranks and seniority, but gave no information about their postings. The second, which was necessary for administrative purposes, continued to detail stations, the units at them, the officers attached to them for administrative and command purposes and their postal addresses. This second list was published and circulated on a strictly controlled basis. If you have a service record for someone who served during Second World War it will usually list stations or units (often simply by number and/or initials) and the RAF Confidential List is an ideal way of tracking down what the unit was, and where it was based. Appendix 12 gives a list of many of the abbreviations used by the RAF and in conjunction with the Confidential Lists (held in TNA in their AIR 10 series between AIR 10/3814 (The Air Ministry Confidential Air Force List April–May 1939) and AIR 10/5421 (The Air Ministry Confidential Air Force List July–December 1953).

The list is divided into commands and then groups, and each group lists the squadrons and units that are part of it. The Fighter Command and Bomber Command Orders of Battle in Appendices 5 and 6 are based on information drawn from the Confidential Lists. For each group, in addition to the squadrons and flights that comprise it, there are lists of the various air staff, administrative and service officers who are employed. No. 3 (Bomber) Group in May 1942 is shown as being commanded by Air Vice Marshal J E A Baldwin, CB, CBE, DSO, with Group Captain J A Grey DFC as his Senior Air Staff Officer and Wing Commander H E Hills as his Senior Administrative Staff Officer. There are twenty named Air Staff officers listed and ten Administrative officers, along with named Navigation, Engineering, Signals, Armament, Photography, Equipment, Accounts and Medical Officers. The Group Chaplain was the Revd J F Fox MC, the Meteorological Officer was Squadron Leader R C Sutcliffe B.Sc. and Flight Officer H C Gray was the WAAF Staff Officer.

Home Units (units based in the UK) are listed geographically so that RAF Belfast can be seen to comprise the RAF Station HQ; HQ RAF Northern Ireland; HQ No. 82 Group; No. 968 Squadron; Queens University Air Squadron; No. 11 RAF Embarkation Unit; Northern Ireland Recruiting HQ; the RAF Movement Control Officer; No. 2 Heavy Mobile W/T Section; the Superintending Engineer, No. 16 Works Area; the Section Officer AMWD and No. 8 Ferry Pilots Pool. There are details of each postal address, telegraphic address and telephone numbers and details of local railway stations.

Overseas Units are also listed geographically, both by commands and by type of unit so that we can see that Helwan (Egypt) houses the Station HQ; Code and Cypher School; No. 1411 Flight (Met); Film Production Unit; No. 2 P R Unit, British Airways Repair Unit; Command Medical Board; No. 1 Australian Air Ambulance Unit and No. 22 Sector Operations Room.

Various Medical Units are listed, including the RAF Hospitals, Convalescent Depots, WAAF Convalescent Depots, the Hospital Base Accounts Office and Medical Stores Depot. The Miscellaneous Units are listed alphabetically, starting with the Aeroplane Armament Experimental Establishment at Boscombe Down and the Airborne Forces Experimental Establishment at Ringway, and including various Flying Boat Repair Stations, Mobile Torpedo Bases, Packed Aircraft Transit Pools, Photographic Reconnaissance Units and the Inspectorate of Recruiting.

George V (in RAF uniform) and Queen Elizabeth, with senior US airmen at an airbase 'somewhere in the Midlands', 1942. (Mrs R J Bone via Mrs R Horrell)

If you have a set of initials on a service record that do not appear in Appendix 12 then find the relevant Confidential List index and plough through it – you should be able to identify it.

Air 29 Records: Miscellaneous Units

In addition to the front line squadrons and their immediate support units and bases, the RAF had an enormous administrative and support organization, each unit of which was expected to keep their own Operations Record Book as if they were an actual squadron. These ORBs are held in AIR 29 Class at TNA.

Simply browsing the AIR 29 catalogue at TNA gives an idea of the number of units involved. A more or less random selection of them include: Ground Control Interception Stations, Sector Operations Rooms, Beach Balloon Units, Bomb Disposal Squadrons, Balloon Centre, RAF Regiment Squadrons, Photographic Interpretation Units, Signal Centres, Photographic Reconnaissance Units, Marine and Launch Sections, Navy and Army Co-operation Units, Anti-Aircraft Practice Camps, Servicing Commandos, Air Sea Rescue Units and Ferry Units (Aircraft Delivery).

Simply in terms of training schools the scale of the RAF during the war is huge. Listings include Parachute Schools, three Schools of Hygiene (two of which were in the Middle East where hygiene was vital), Physical Training Schools, Schools of Ship and Aircraft Recognition, Radio Schools, a Glider Instructors School, Schools of Technical Training, Officers Schools and the RAF Chaplains School at Cambridge.

The index for AIR 29 also shows the incredible importance of signals in the Second World War. There are hundreds of references to Signal Centres, Mobile Signals Units, Heavy Mobile W/T Units and Air Transportable Signals Units. Almost all are designated by numbers, though a few have place names. AIR 29/156, for example, contains ORBs for Nos. 10 and 12 Signals Unit as well as Signals Station, Freetown, and RAF Signals Unit, Tezpur.

One of the problems with trying to read a service record is that most postings are referred to by abbreviations, partly because of the service's love of acronyms, but mainly because of the need to keep references short for reasons of space. There is an attempt to set out many of the more common abbreviations, and to explain how you can interpret more, in Appendix 12.

A Miscellaneous Unit: the RAF Intelligence School

The RAF School of Intelligence (RAF Highgate) was established on 1 September 1942 in four requisitioned properties. The main one was Caen Wood Towers in Hampstead Lane, London, which had been requisitioned from Sir Robert Waley-Cohen, and which served as the mess and accommodation for the permanent officer staff and officers attending courses, as well as providing lecture rooms. Numbers 7, 13 and 19 Sheldon Way were also requisitioned and served as the other ranks mess and accommodation for men and women other ranks respectively. 'The function of the unit is the instruction of Officers in Air Intelligence subjects and the assessment of new candidates for Intelligence duties in Operational Commands and Air Ministry Departments.' The school ran three types of course: (A) which was to provide a broad and general background to air intelligence for officers with some experience, (B) for more senior officers dealt with escape and evasion (and was arranged by the secret organization MI9) and (C) for newer officers intended for air intelligence duties so that they could be assessed and allotted roles to suit their particular qualities and expertise.

The station came formally under RAF Training Command through HQ 27 (Training) Group but accounting, equipment, medical and dental services were undertaken by RAF Hendon of 44 Group Ferry Command.

The ORB (AIR 29/715) contains numerous lists of personnel and their duties, particularly those of the various officer instructors, though many NCOs and other ranks, including WAAFs, get mentioned in the various rotas which cover fire duties, station NCO and clerks. If a relative served at the school there is a good chance of catching their name on one or other of the Station Daily Routine Orders which the ORB contains. There are also Personnel Occurrence Reports which give details of officers joining or leaving the station, so that we see Pilot Officers G Bettelley and E G Evans posted to HQ Bomber Command, Pilots Officers S I Williams, A D Maclaren and A H Daoust to HQ Fighter Command and A F M Marks and H R Everett to HQ Coastal Command, all on 2 December 1942.

Most of the daily routine of the station consists of the various courses passing through, and unfortunately there isn't a record of the men (and women) who attended them. Periodically the ORB records visits by senior officers and the regular Anti Gas Drills and fire practices but twice there are reports of a more serious nature. At 07.30 hours on 16 June 1944 a pilotless aircraft (a V1) landed and exploded some 88 yards from the station, causing superficial damage by blowing in four large windows. No one was hurt and the station's operational capacity remained unaffected. Twelve days later, on 28 June 1944, another V1 landed close to the station, injuring three airmen slightly and doing extensive damage to Caen Wood Towers and the Guard Room. Approximately 312 windows and skylights were blown in and some tiles stripped from the roof. The Maintenance Department of the Ministry of Works attended within 30 minutes of the blast and began repairs and Squadron Leader Parsons, the Commanding Officer, was able to report that operational capacity had again not been affected.

There are various course syllabuses scattered throughout the ORB, which also name the course instructors. Escape and evasion for aircrew who might be shot down are examined in courses run in conjunction with MI9, the top secret organization set up to help escapees. Major Airey Neave of MI9 (famous as an escaper from Colditz) gave a lecture on 'This War Experiences', there were talks on gadgets (presumably MI9's miniaturized compasses and other escape materials) and an American film *Resisting Interrogation*. A course given at the end of January 1945 concentrates on the war with Japan with lectures on (amongst others) 'Japanese Air Forces – Organisation, training and strength', 'Bushido', 'Topography of Burma', 'Japanese Psychology' and 'The Japanese Character in War'. Films shown included *Japan the Enemy*, *Castaway* and *The Last Enemy*.

Second World War prisoners of war

Many thousands of Air Force personnel were captured by the enemy during the war. Over the last few years an increasing number of records relating to them have been released to TNA.

As the war came to an end MI9 carried out a mass screening of all prisoners of war. Preprinted questionnaires were given to all prisoners, and the completed results are in WO 344. Even though this is a War Office series, it contains the reports of captured air and ground crew and the questionnaires are held alphabetically so you don't need a service number to find the person you are interested in. The questionnaires ask about place and date of original capture, camps held in, any illnesses suffered, adequacy of medical treatment, interrogations, training received on escape and evasion, escapes attempted, sabotage undertaken, knowledge or suspicion of any collaboration with the enemy and knowledge of any war crimes.

Warrant Officer Thomas Geoffrey Brook, captured at Haarlem on 26 June 1943, gave details of the three camps he'd been held at, confirmed he'd received training at RAF Lindholme on how to behave in the event of capture, and that he had attempted to escape twice. At Fallingsbostel Station on 6 April 1945 he'd escaped with Warrant Officer Webster but was captured by the SS. Escaping for a second time he was recaptured by the German Home Guard but later freed by the British Army.

Flight Sergeant John Edward Bromley was captured at Dinslaken near Krefeld on 22 June 1943 and held at four camps. At Dulag Camp, where he was initially held, he says that he was held for four days in solitary confinement without exercise or fresh air and threatened with death for not revealing secrets. He escaped from a marching column of prisoners with Warrant Officer Tout and managed to hide but, being recaptured after a few hours, they were attached to a sick column when it was discovered they were ill.

Other MI9 POW reports can be found in the AIR 40 (Intelligence) papers and often give more detailed reports than the standard interrogation papers as they were compiled from debriefing of escapees. Flight Lieutenant H J Turner describes (AIR 40/1533) how he was shot down in his Hurricane of 33 Squadron on a ground attack raid against targets behind enemy lines in Egypt on 3 November 1942. Having been wounded (and threatened by his Italian captors, from whom he was 'rescued' by German troops) he was sent eventually to a camp in Italy at Fomtellato. Here he helped the escape committee by making escape equipment while making scenery for the camp theatre.

After Italy's surrender in 1943, with the connivance of the Camp Commandant, the entire camp marched out and went into hiding to avoid the local German garrison. After two days the men broke up into groups and were ordered to make their own way either to British lines or Switzerland. Turner's group made their way north and, with the aid of local Italians, avoided the Germans and escaped into Switzerland in January 1945.

Flight Lieutenant A B Thompson of 102 Squadron was one of the lucky escapers who took part in 'The Great Escape' from Stalag Luft III in March 1944. The escape was later made famous in the film of the same name starring Richard Attenborough and Steve McQueen. Thompson was actually shot down on 8 September 1939 and, as one of the first RAF prisoners of war, was interrogated by Goering himself. In March 1944 he was the sixty-eighth person out of his escape tunnel and made his way towards the nearby railway but was captured by civilian police. After ten days in a prison cell, where he was threatened with death, he was returned to a POW camp. Only three men escaped to Britain; nine were discovered in concentration camps after the war. Fifty escapees were shot and Thompson provided complete evidence of this to the Swiss Protecting Power (who tried to safeguard prisoners' interests) and gave evidence to the Judge Advocate's Office and War Crimes Commission after he was liberated by British troops in May 1945.

Documents in the AIR 40 section of the TNA covering the Great Escape are extremely comprehensive and contain many documents relating to, or even signed by, men who either escaped and survived or who conducted the investigations. AIR 40/268 is the official history of Stalag Luft III, and other documents in the range AIR 40/2265 to AIR 40/2293 refer in great detail to the investigation and its progress as well as giving details of camp life, such as correspondence between the Kommandant and the SBO and many preserved internal documents, all of which are well worth a browse.

An excellent website is http://www.elsham.pwp.blueyonder.co.uk/gt_esc. This has photographs of the escapers and many of their captors and the Gestapo officers who pursued them. It also recounts in detail the story of the escape, the murders, the subsequent RAF investigation (by the RAF Special Investigation Branch) and trials of the killers.

Other MI9 files specific to RAF POWs and escapees are in AIR 40/1932 and AIR 40/1933 and other files relating to the organization can be found in AIR 40/2450 and 2451 covering the Far East, and HS 7/172 and 173 which cover the Eastern Mediterranean. Other files in WO 208 include details of the organization, some staff files and copies of reports.

There are a few nominal rolls of prisoners held by the Germans in the AIR 40 series. AIR 40/263 and 264 are nominal rolls of prisoners held in Stalag Lufts I and IV. AIR 40/269 lists prisoners in Stalag Luft III; AIR 40/270 is from Stalag IIIA; AIR 40/272 is Stalag 17B; AIR 40/276 and 277 Stalag 357. AIR 40 1488 to 1491 cover Stalag Luft III and there is a roll of other ranks prisoners for repatriation.

Prisoners of the Japanese

The National Archive holds record cards for British prisoners of war in the Far East in their WO 345 series, which consists of some 56,000 preprinted cards which were compiled by an unknown central Japanese authority with some degree of Allied assistance.

The cards record the following details:

- camp (in Japanese);
- name;
- nationality (in English, or English and Japanese);
- rank (in English, or English and Japanese);
- place of capture (in English, or English and Japanese);
- father's name;
- place of origin;
- destination of report (assumed to be report of capture, sent to next of kin at address given);
- prisoner's camp number (in Japanese);
- date of birth;
- unit and service number;

- date captured (in English, or English and Japanese);
- mother's name;
- occupation (in English, or English and Japanese);
- remarks;
- other information (on reverse) which may include medical details (in Japanese, possibly partly translated).

Diagonal red lines across a card indicate that the prisoner died in captivity.

Please note that dates are written in the American style, i.e. 31 December 1918 is written 1918.12.31 and that the year of date of capture is given as 16, 17, 18 or 19 which seem to be 1941, 42, 43 and 44 respectively.

An aircraftsman's record card

The record card for 1205956 Reginald Arthur Heath shows that he was an Aircraftsman Second Class of 605 Squadron captured on 17 March 1942. Records for 605 Squadron for the period are sparse – they were sent out to the Far East on the outbreak of war with Japan and were soon caught up in the debacle that led to, and followed, the capture of Singapore. The official history of the squadron on the RAF website says:

> It reached Singapore in January 1942, too late to affect the campaign, and was evacuated to Sumatra on arrival in the area, moving later to Java. There it became caught up in the Japanese invasion and after operating a collection of surviving aircraft, was either evacuated in small groups or captured by the Japanese by early March.

The Operation Record Book for 605 Squadron covering the period is in AIR 27/2089. On 2 October 1941 the ORB says:

> We are posted overseas. The news is received with great jubilation. No details are known, but the 25th of this month is given as the date.

Aircraftman 2nd Class Reginald Arthur Heath RAFVR. (Courtesy of Nick and Carmen Coombs)

> Rumours of every description are circulating. The Squadron becomes non-operational tomorrow. Is it Leningrad, Singapore, Teheran or Cairo?

After many changes of plan, cancellations and rumours the final note for 31 October 1941 reads:

> Our overseas establishment is now complete, and consists of 22 officers, 3 W/O's and 308 NCO's and airmen. We now enthusiastically look forward to whatever November may bring.

A detachment of 605 Hurricanes was left in Malta en route to the Far East and a couple of pages of the ORB relate to them, but the next entry, for 7 June 1942, is particularly poignant:

> 605 Squadron commenced to re-form. The Squadron lost its identity when it proceeded overseas in the early part of the year. It is believed that most of the ground personnel were taken prisoner in Batavia.

AIR 23/2123 contains a report by Air Vice Marshal P C Maltby on 'Operations of the R.A.F. during the campaigns in Malaya and Netherlands East Indies 1941–1942' which gives some brief explanation of how Reginald was captured. 232 and 605 Squadrons were based together as 266 (F) Wing at Tjililitan for the protection of Batavia. 605 Squadron 'had hitherto been mainly employed in erecting Hurricanes for No. 266 (F) Wing as a whole, but it was not armed until it came into action on 23rd February with 8 Hurricanes'. The Japanese were making landings all along the Javanese coast and 'From 17th to 27th February this force was continually in action in its role of the air defence of Batavia. Normal odds met were in the vicinity of 10–1.' It was hoped that an American Aircraft Carrier, USS *Langley* might bring in new P40 fighters to re-equip the squadrons, but she was sunk on 27 February by Japanese bombers and it was decided to evacuate 605 Squadron but to ask for volunteers to stay behind to help 232 Squadron. We can assume that Reginald was one of these volunteers.

232 Squadron now found itself attacking Japanese landings along the coast, destroying three Japanese flying boats and shooting up columns of advancing troops. By 2 March their airfield was under constant aerial attack and Japanese troops were closing in. Under orders from Group Captain Noble they moved to a base near Bandbeng. All the time it was engaged in battle with superior Japanese numbers in the air. By 5 March the Dutch High Command were advising that further resistance was becoming impossible and the various British units, including the remnants of 605 Squadron, took to the hills to continue resistance as long as the Dutch continued to fight. The Dutch Commander, having obtained a Japanese assurance that prisoners of war would be treated in accordance with the Geneva Convention, signed the surrender. The RAF managed to destroy most of their equipment and, along with the remainder of the British forces, were mustered by the Japanese and marched to Batavia.

> The Japanese subsequently endeavoured to extract information from aircrews of Nos. 232 (F) Squadron . . . and from other individual officers and airmen with almost complete lack of success, in spite of protracted brutal treatment in many cases . . . The later treatment of P.O.W's with little regard to the terms of the Geneva Convention . . . is too well known to need further elaboration in this report.

收容所 Camp	爪哇 昭17年 8月 1日	番號 No.	爪本 4429
姓名 Name	Heath, Reginald Arthur. ヘース、レヂナルド、アーサー。	生年月日 Date of Birth	1910. 6. 5.
國籍 Nationality	英	所屬部隊 Unit	No. 1205956 Royal Air Force. 1205956X 605 Sqiadron.
階級身分 Rank	Aircraftsman 2. 空軍上等兵		
捕獲場所 Place of Capture	爪哇	捕獲年月日 Date of Capture	昭和 17 年 3 月 20 日
父ノ名 Father's Name	Heath, William.	母ノ名 Mother's Name	Heath, Ellen.
本籍地 Place of Origin	Frome, England.	職業 Occupation	Painter & Decorator. ペンキ塗屋.
通報先 Destination of Report	Wife: 6 The Ha[illegible]s, Fromfield, Frome, Somerset, England.	特記事項 Remark	

Reggie Heath's Japanese prisoner of war record card. (TNA WO 345/24)

Reg Heath surrendered to the Japanese with the main body of the RAF on 17 March and was held in a POW camp on the small Indonesian island of Ambon. In November 1943 he was put aboard ship to be transported back to Java, along with 414 other Britons and 133 Dutch prisoners, many considered too weak to work. His ship, the *Suez Maru*, was attacked by an American submarine USS *Bonefish* and sunk on 29 November 1943 and Reg was among many of the prisoners who drowned. Author Allan Jones, whose father was among those killed, has shown in his book *The Suez Maru Atrocity: Justice Denied* (Allan C Jones, 2002) that, though some prisoners were drowned in the ship, some 200 were machine-gunned in the water by the Japanese navy.

Reginald Arthur Heath is remembered on the Singapore Memorial to the Missing. His record card is scored through with the red line that denoted his death.

Other prisoner of war records

WO 224 series comprises a collection of reports by the International Red Cross and Protecting Powers (Geneva) and deals with conditions and events in various prisoner of war camps in Europe and the Far East. RAF prisoners of the Germans were held mainly in Luftwaffe prisoner of war camps, hence names such as Stalag Luft III. The reports detail conditions in the camps, talk about rations, cleanliness, attitudes of the guards, numbers held, mail and entertainments. WO 224/62 covers Stalag Luft I at Barth in eastern Germany between June 1941 and April 1945.

Though individuals are rarely named unless they hold some position within the camp, there are occasional lists of names of men in hospital, with details of their ailments.

http://www.merkki.com is an American site created by the children of an American flyer who was held at Barth and is well worth a look even if your relative was at a different camp as it shows the amount of information that can be gathered given enough time and effort.

There is Nominal Roll of RAF prisoners held at the notorious Changi camp in Singapore in AIR 40, between AIR 40/1899 and AIR 40/1906.

Prisoner of war record cards for men captured by the Germans and Italians are still held by the Ministry of Defence but a copy of an individual's card (subject to the standard release only to next of kin or the man himself) can be obtained from the Veterans Agency in Blackpool. The card is similar in design to the Japanese cards in TNA, giving very much the same kind of information, including date and place of capture, the camps held in, name and address of next of kin.

Log books

If you're lucky enough to have a relative's log book, it can save you a lot of work as there's a great deal of information contained in it. Everyone who flew regularly was required to keep a log book and to keep it up to date. Partly this was because you were entitled to flight pay depending upon a minimum number of hours flown each month. The log book had to be signed off monthly by your senior officer. Even without a service record the log book should tell you which squadron the owner was with, which bases they flew from, their aircraft and the type and duration of their flights.

It wasn't only pilots who kept a log book. Observers and air gunners did too, and RAF log books were also used (in the Second World War) by glider pilots, Fleet Air Arm pilots and crew and the Army Air Corps.

From the very earliest days of aviation the information contained in the log book has remained pretty much unchanged. AIR 1/1625/204/89/2 contains the log book of Captain J M Salmond, one of the Central Flying School's instructors, covering May 1913 to July 1914. As well as listing the aircraft that he flew, it lists the duration of each flight, his passengers (usually trainee pilots), the destination of any flight away from the airfield and remarks on any incidents. On 13 June 1913 Salmond was training Sergeant Rigby in a BE aircraft number 417 when Rigby 'lost flying speed in the air, failed to depress control lever, slipped and broke machine'. As well as detailing flights that Salmond himself took it also lists some of the individual flights made by his pupils, so that on 9 June 1913 we can see that Lieutenant Newall, as well as being a passenger in a 20 minute flight,'did straights and circuits alone, 20 minutes'.

A sample range of log books dating from 1916 to 1975, and taking in most of the RAF's operations overseas, as well as the Battle of Britain and major Bomber Command operations, are in TNA's AIR 4 series.

Using a flight sergeant's log book to find more information

The log book of Flight Sergeant Ronald Cooke (1802970), who served as an Air Navigator, shows that he qualified with effect from 10 July 1943 at 41 Air School,

East London, South Africa. He passed the training course with a final mark of 74 per cent and was described as 'Capable and Reliable'. The log book details the various flights that he made as part of his training, along with the names of the various pilots he flew with.

On his return to the UK he was posted to No. 62 Operational Training Unit (OTU) at RAF Ouston in September 1943 to train in radio navigation. Over the next few months he trained with No. 51 OTU, No. 63 OTU at RAF Honliley and was posted to 151 Squadron flying in Mosquitos on 20 May 1944.

The ORBs for 151 Squadron covering the period that Ron flew with them are between AIR 27/1022 (1944 Jan.–1945 May) and AIR 27/1023 (June 1945–Sept. 1946), with 151 Squadron Combat Reports in AIR 50/63, though it is obvious from the log book that, though Ron flew sorties and saw action, he did not take part in any air fighting.

According to AIR 27/1023, part of the role for 151 Squadron was night fighter defence but on 18 July 1944 the ORB records

> WITH EFFECT FROM TODAY, 18th JULY 1944 THE NIGHT FIGHTER DEFENSIVE ROLE OF No. 151 SQUADRON IS TO CEASE UNTIL FURTHER NOTICE. No. 151 SQUADRON, WHICH WILL BE WHOLLY EQUIPPED WITH MOSQUITO VI FIGHTER BOMBERS WILL BE

Time carried forward :— 176.05 47.20

Date	Hour	Aircraft Type and No	Pilot	Duty	Remarks (including results of bombing gunnery exercises etc.)	Flying Times Day	Night
13.7.44	1505	Mosquito HK462	F/O Cox	Nav. Radio.	N.F.T.	.25	
14.7.44	1030	Wellington 534	F/O Collier	Nav. Radio.	Mk. 10. Practice	.55	
14.7.44	1515	Wellington 524	F/L Smith	"	"	1.00	
17.7.44	1445	Wellington 534	F/O Collier	"	"	2.00	
18.7.44	1025	Wellington 526	F/O Cook	"	"	1.50	
19.7.44	1415	Wellington 526	F/L Smith	"	"	2.10	
19.7.44	1815	Wellington 526	F/L Smith	"	"	2.00	
20.7.44	2100	Wellington 548	F/L Smith	"	"	1.20	
22.7.44	1030	Wellington 548	F/L Smith	"	"	1.40	
22.7.44	1540	Mosquito NT194	W/O Prichard	Navigator	Ranger.- Bomb Rlwy, Chateaurenault - Tours	3.35	
24.7.44	1505	Mosquito PZ196	W/O Prichard	"	Air Firing, Cross-Country.	1.40	
26.7.44	1530	Mosquito PZ186	W/O Prichard	"	Ranger - Bomb Rlwy Near Nantes.	3.00	
31.7.44	2100	Mosquito NS935	W/O Prichard	"	Patrol - Naval Escort South Of Brest Peninsula	1.00	2.3[illegible]

Summary for July		Day	Night
Unit 151 Sqdn.	Mosquito	9.40	2.35
July 31st 1944	Wellington	12.55	–
Sig. R.J. Cooke	Total	22.35	2.35

W. E. Thomas F/Lt. Nav. Rad. Leader

Total Time — 198.40 49.55

Excerpt from the log book of Flight Sergeant Ronald Cooke from July 1944 showing attacks on the French railway system and various training flights. (With thanks to Joyce Cooke and family)

Observer (but not yet Flight Sergeant) Ron Cooke after his return from training in South Africa. (With thanks to Joyce Cooke and family)

EMPLOYED IN AN OFFENSIVE ROLE INCLUDING BOMBER SUPPORT.

On 7 August 1944 the log book shows that Ron flew with Warrant Officer Prichard on a 'Ranger' flight over France. These were flights to attack targets of opportunity, and with the Battle of Normandy in full flow, targets were likely to be numerous as the Germans moved men and equipment to the battlefield. The log book says simply 'Ranger. Bomb Niort-Poitiers Rlwy Line' but the ORB gives a little more detail: '7 August 1944: Bombing Rangers by F/O Turner and F/O Partridge with W/O Prichard and F/S Cooke . . . Railways and a Barracks were bombed.'

Other Ranger flights, mainly bombing railways, are recorded, but later in the year the squadron went back to a night fighter role. Part of this involved supporting Bomber Command raids by flying over German airfields in the hope of catching German night fighters when they were most vulnerable, landing and taking off. The log for 21 February 1945 records 'Bomber Support – Hanover, Targets Duisberg, Worms', but again the ORB gives more detail:

INTRUDER PATROL TO WUNSDORF, 21/22nd FEBRUARY '45

One Mosquito MK XXX A I MK X (W/O Prichard and F/S Cooke) was a/b from Hunsdon at 1905 hours on Intruder patrol to Wunsdorf. Crossed out English coast at Orfordness at 1928 hours. In Continental coast at 4,000' at Egmond at 1955 hours. Saw white beacon 'B' at 2015 hours position 5245N 0620E and further one flashing 'AKE' at 2039 hours position 5300N 0825E. Reached patrol at 2045 hours almost immediately coned in s/l's but no flak. This was repeated 4 or 5 times during patrol. A/f was not lit up. Left patrol at 2200 hours and out Dutch coast between Terschelling and Vlieland at 2237 hours. In English coast at Orfordness and landed at Bradwell Bay at 1125 hours owing to poor visibility at Hunsdon. Weather : 7/10ths at 15,000' 2/10ths low cloud. Visibility good over land deteriorating on return.

As the war drew to a close more bomber support operations are recorded, with night operations against Hamburg, Celle, Nuremberg and Grove. After VE Day the squadron settled down to training and routine work, but in June 1946, based at RAF

Group photo of C Flight, 3 Squadron, No. 3 Initial Training Wing, South Africa. Ron Cooke is fourth from right in top row. (With thanks to Joyce Cooke and family).

Exeter, Ron was chosen to be one of four aircrew to take part in the Victory Day parade over central London. With Flight Lieutenant Spencer he flew to RAF Tangmere on 2 June, flew practice flights on 3, 6 and 7 June, then, on 8 June 1946: 'Bigbull Victory Parade Fly-Past over London'.

The ORB is slightly more descriptive:

> TANGMERE 8 June: V-DAY. Those taking part in the Fly Past were up early and casting anxious glances at the weather which did not appear to be as good as Met had forecast. Our four aircraft in the combined 85/151 Squadron together with our one reserve were airborne and formed up to time. The route was via LUTON and STOWMARKET, where we joined the Mosquito Stream, and thence down to BRADWELL BAY and on to FAIRLOP, the starting place for the run over LONDON for all the 300 or so aircraft taking part. From this point onwards the weather deteriorated and over the saluting base itself in The Mall it was pouring with rain and there was little to be seen. However, after a circuit of NORTH LONDON, the Squadron made its way back to TANGMERE, Unfortunately, the formation found itself in ten tenths cloud and had to split up immediately. Arriving over TANGMERE to find a heavy thunderstorm in progress and cloud base down almost to the deck, hasty landings were made, one at TANGMERE by F/L SPENCER . . . Subsequently the weather improved and all the aircraft with the exception of one left behind at Ford for fabric repairs were able to return to EXETER in time for the crews to take part in the evening Victory celebrations and go away for the Whitsun Weekend.

151 Squadron at Exeter in front of Mosquito, 1945. (With thanks to Joyce Cooke and family)

The log book ends with Ron and Flight Lieutenant Spencer flying back to Exeter. The final log book entry is a note by the Officer Commanding 151 Squadron dated 3 July 1946, wishing Ron 'the best of luck in the street'.

Other files relating to Ron's service, some of which are very informative, others less so are:

AIR 54/171: No. 41 Air School 1940 Sept.–1943 Oct.

AIR 28/624: OUSTON 1941 Mar.–1943 July

AIR 16/784: No. 62 OTU 1942 Aug.–1944 Aug.

AIR 28/369: HONILEY 1940 Sept.–1945 Jan.

AIR 29/685: No. 63 Honiley 1943 Sept.–1944 Feb.

AIR 2/6928: Victory march, London, 8th June 1946: arrangements

AIR 2/9730: Victory Day Celebrations June 1946: air participation

AIR 16/1158: No. 11 Group Order for Victory Day Fly-past and Amendment List No. 1 1946 Apr.–May

Other RAF personnel

RAF nurses

The Royal Air Force Nursing Service was founded on 1 June 1918 and established as a permanent part of the RAF by Royal Charter in January 1921. Princess Mary became royal patron in June that year and the service was renamed Princess Mary's Royal Air Force Nursing Service. The first RAF nurses went to serve in Iraq in 1922.

During Second World War RAF nurses served in every theatre of the war. By 1943 there were thirty-one RAF Hospitals and seventy-one Station Sick Quarters.

Papers relating to RAF Medical Services during Second World War are in TNA's AIR 49 series. Though these are usually compiled on a command or district level, they often contain reports from individual RAF medical units.

The Operation Record Books for the hospitals and other medical units are in AIR 29.

http://www.pmrafns.org/index.htm, the excellent website for the service, gives a brief history of the service, along with details of the current service and careers for RAF nurses.

The Women's Auxiliary Air Force

The Women's Auxiliary Air Force was formed on 28 June 1939. Though in the First World War most women had been given domestic and clerking jobs, it was always intended that the WAAF would work in such front line posts as radar operators and working in sector control. As the war continued, WAAF servicewomen worked in meteorology, transport, telephony and telegraphy, codes and ciphers, intelligence, security and operation rooms.

There are numerous files on women's service in the Air Force in the AIR 2 series detailing pay, conditions and some of the perceived problems of having women serving alongside men. On the whole WAAFs were considered to have done a

The arrival of the first WAAF contingent at RAF Pembroke Dock, 1940. (Mrs R J Bone via Mrs R Horrell)

WAAFs working on a flying boat at RAF Pembroke Dock. (Mrs R J Bone via Mrs R Horrell)

splendid job and in 1949 the WAAF was reformed as the Women's Royal Air Force and was amalgamated with the RAF in 1994.

Service records for WAAFs and WRAFs are still held by the Ministry of Defence so if seeking them you'll need to write to RAF Innsworth. The service record will tell you the units your relative was posted to and you can search for unit records in the relevant AIR series. Group and wing records are again in AIR 25 and AIR 26 and AIR 27, AIR 28 and AIR 29 squadrons, stations and miscellaneous units respectively. Foreign Command records are in AIR 24.

http://www.waafassociation.org.uk is an excellent website devoted to the WAAF.

The RAF Regiment and earlier ground forces

The Royal Air Force Regiment was formed in 1942 in response to problems the RAF had suffered since the beginning of the war in protecting their airfields. Heavy losses had been suffered in France in 1940 when retreating units had been overrun. During the Battle if Britain it was felt there was insufficient anti-aircraft defence and in Crete in 1941 the principal RAF base at Malame was overrun by German paratroops with little defence being offered.

The new RAF Regiment was composed of anti-aircraft flights and field squadrons, but all members of the regiment were initially trained as infantrymen. The regiment absorbed the RAF armoured car units that had served in Iraq since 1921. Some of the earliest RAF armoured car records are in AIR 1. AIR

1/431/15/260/18 comprises various reconnaissance reports from Iraq between October 1921 and October 1922. AIR 5/838 to AIR 5/841 are unit War Diaries for the Armoured Car Companies 1923–4. AIR 29 holds the ORBs for the Armoured Car Companies from 1927 onwards.

AIR 29/50 includes the ORB for the No. 1 Armoured Car Company which took part in the dramatic defence of RAF Habbaniya in 1941 when it was besieged by Iraqi forces. The ORB notes, at 05.00 hrs on 30 April 1941: 'Iraqi Forces invested Cantonment and issued ultimatum to cease flying or shelling would commence. The ultimatum was rejected. The Company stood by at instant notice.'

The ORB is brief in its daily description of operations over the next few days as the hastily gathered force of converted training aircraft and antiquated second line aircraft attacked the surrounding Iraqi forces. It does, however, give details of individual incidents separately. On 2 May:

> No. 3 Section made a Sortie with a view to attacking six enemy armoured cars in position on the south side of the aerodrome. Five attacks were made and fire was opened at approximately 400 yards range, with armour piercing machine gun bullets, scoring hits on the enemy but not disabling any of them. All our vehicles returned safely. There were no casualties but several punctures were caused by enemy fire.

A full narrative of No. 1 Armoured Car Company's part in the defence is in AIR 23/5982.

ORBs for the RAF Regiment are in AIR 29: AIR 29/883 covers No. 4010 Anti-Aircraft Flight at RAF Skitten, near Wick in Scotland. You'll need to be aware that searching for individual units can be slightly complicated – some units are indexed under 'Regiment', others under 'Regt' or under 'Field Squadron'.

Other flying services in the Second World War

The Fleet Air Arm

The Fleet Air Arm was busy from the start of the Second World War. *Ark Royal*'s fighters shot down the first German aircraft of the war and a Sea Fox from HMS *Ajax* spotted for the cruisers at the Battle of the River Plate. When Germany invaded Norway in 1940 FAA aircraft dive-bombed and sank German cruiser *Konigsberg* in Bergen harbour. Unfortunately HMS *Glorious* was later surprised by the *Scharnhorst* and sunk before she could launch her aircraft.

During the Battle of Britain twenty-two FAA pilots were rapidly retrained on Spitfires and Hurricanes and loaned to the hard-pressed RAF. FAA squadrons 804 and 808 took part in their Skua and Fulmar fighters. Nine FAA pilots were killed during the battle.

In the Mediterranean, twenty-one Fairey Swordfish biplanes attacked the Italian Battle Fleet in its home port of Taranto. Three battleships were torpedoed and severely damaged, for the loss of two men killed and two captured. In 1941 FAA aircraft damaged the Italian cruiser *Pola*, leading to the successful Battle of Matapan. Naval aircraft from Malta attacked Axis convoys and, in 1942, FAA aircraft supported the convoy for Operation Pedestal (the relief of Malta), though HMS *Eagle* was sunk by a U-boat and other carriers damaged by German aircraft.

In the North Atlantic, in the hunt for the *Bismarck*, both HMS *Victorious* and *Ark*

Royal launched air strikes at the battleship. One from *Ark Royal* finally hit *Bismarck*'s steering gear leaving her unable to escape the British fleet, which sank her with guns and torpedoes. In 1942 FAA Swordfish flew against the German warships *Scharnhorst*, *Gneisenau* and *Prinz Eugen* when they dashed through the English Channel. Led by Lieutenant Commander Eugene Esmonde they pressed home an attack against heavy anti-aircraft fire and in the face of German fighters. All the Swordfish were shot down. but Esmonde managed to drop his torpedo, in spite of his aircraft being badly damaged, and for his valour was awarded a posthumous Victoria Cross.

In the North Atlantic the FAA flew anti-submarine operations for convoys. Initially aircraft were flown off catapult aircraft merchantmen which launched Sea Hurricanes (with rocket assistance) from a rail. The plane couldn't land except by ditching in the sea, with the pilot hopefully being picked up. Later, merchant aircraft carriers were introduced – merchant ships with wooden flight decks laid over their normal decks. Though difficult to fly from they at least did give the option of a safe landing for the aircraft coming back from patrol. With the aid of their air cover the battle against the U-boat was gradually won.

FAA pilots and carriers served in the Indian and Pacific Oceans against the Japanese. HMS *Hermes* was sunk off Ceylon in 1942. In 1945 FAA planes carried out their largest ever air raid against oil refineries at Palembang on Sumatra and other carriers supported American operations against Okinawa and the Japanese mainland, until her surrender in August 1945.

By 1945 the FAA had fifty-two operational carriers with 3,243 pilots.

Fleet Air Arm: 804 Squadron

No. 804 Squadron was formed as a shore-based fighter squadron on 30 November 1939 to counter enemy air activity over Scapa Flow, main base of the British Home Fleet. It was one of two FAA squadrons credited with taking part in the Battle of Britain. Its squadron diary covering September 1939 to December 1944 is in ADM 207/8 and contains a nominal roll of officers and other ranks, including the ratings' numbers and trades. FAA diaries tend to be quite chatty, which gives a good feel for what was going on, but tend to be lacking in the detail of individual flights that you'll see in an RAF ORB:

> [Monday 8 April 1940] A day of peace until 2000 when the fun began as the sector prepared a welcome for a large plot approaching from the S.E. This developed into a raid on the ships at Scapa at 2030. Clouds were 7/10 at 2000' except over the Flow, which was clear; wind slight from the North & visibility good except in rain showers which were floating round to the N & NE. At 2000 the Yellow section went up to keep goal over Copinsay at 17,000'; 20 minutes later the syren wailed and at 2025 Red scrambled to 'M', followed 5 minutes later by Green and Blue who patrolled over Hatston and to the North. Although our sections sighted and pursued E/A at times no contact was made. Most of these were either disappearing into clouds or at a very great height so out of reach. There was a general pancake about 2130. The firing and searchlights continued for about 20 minutes longer.
>
> The enemy were coming in in bunches of six apparently and they failed to hit any targets or to do any damage whatsoever. 8 Hurricanes from Wick intercepted one batch of 6 Heinkels well out to sea, of which they shot 4 down. Shortly after they had landed a Heinkel appeared doing steep turns

> around Wick aerodrome. One Hurricane rushed off the ground in pursuit but before he could get into position the Heinkel landed on the flare path. It was really very badly shot about, the 2 rear gunners were dead though the 16 year old pilot and his observer were not hurt.
>
> [Saturday 21 December 1940] Reds were still picking the breakfast out of their teeth when there came a loud shout of Scramble – the news that there was a bandit over the circus. Guns could be seen blazing from the circus – Hatston reported an E/A. Reds rushed to Hatston, saw nothing so shot into the attic where it was a lovely day and there was some stale gunfire floating around. The bastard had made the popeye – his plot became mingled with those of the Red Section (attic) and a Hurricane section (cellar) and was next picked up 30 miles East going like hell for Stavanger.

Between 1940 and 1944 the squadron flew off catapult armed merchantmen on convoys, did escort duty for the North Africa landings and for Atlantic convoys. In April 1944 it took part in a carrier-borne attack on the German battleship *Tirpitz*:

> [3 April 1944] At 0415 the first strike was flown off to attack the German Battleship *Tirpitz* which was lying in the Kaafjord in Norway. An hour later the second strike was flown off, 804 Squadron being in the 2nd strike. Corsairs did top cover while Wildcats and Hellcats strafed gun positions and flak ships. Barracudas did the bombing and 17 definite hits were obtained. 1 Hellcat (800 Squadron) ditched owing to damaged hook. The pilot S/L HOARE (800) was picked up by a destroyer. 3 Barracudas were missing. The fleet then returned to Scapa Flow suffering the usual alarms from 'Bogey' aircraft and U-Boats.

In 1945 the squadron was sent to the Indian Ocean and took part in operations against the Andaman Islands and in the Dutch East Indies before returning to the UK to be disbanded on 18 November 1945.

Fleet Air Arm records 1939–1945

ADM 207 contains ORBs for many Fleet Air Arm squadrons, including 822 Squadron (ADM 207/18), 826 Squadron (ADM 207/22 and ADM 207/23) and 880 Squadron (ADM 207/37). There are also some ORBs in AIR 27/2387.

As well as Squadron Diaries you'll find reports on Fleet Air Arm operations as part of the Admiralty papers on operations that they took part in. There are also a series of Squadron reports in ADM 199/115, ADM 199/166, ADM 199/167, ADM 199/479 and 480 (covering FAA operations in Norway in 1940) and between ADM 199/838 and ADM 199/842. ADM 199/114 contains reports on naval air operations in support of the North Africa campaign and in the Eastern Mediterranean 1941 to 1943 and ADM 199/844 reports on naval air operations off the Norwegian coast in 1944.

ADM 1 papers contain many on the Fleet Air Arm, though most are to do with organization and equipment. A few refer to specific incidents such as

> ADM 1/12229: ACTIONS WITH THE ENEMY (3): Sub-Lieutenant (A) R.C. McKay, RNVR and Acting Leading Airman D.H. Stockman Fleet Air Arm both killed in action at Suda Bay awarded posthumous mentions in despatches;

ADM 1/15761 Preliminary interpretation report on Fleet Air Arm attack on TIRPITZ 1944;

ADM 1/15805 Fleet Air Arm, attacks on enemy submarines by aircraft from HMS STRIKER: reports on attacks made 22/23 August;

ADM 1/16840 Fleet Air Arm attack on Tirpitz: preliminary report on Northern Norway.

A few ADM 1 papers refer to awards of medals to FAA personnel, including:

ADM 1/12381 830: Squadron: awards to personnel engaged in operations in the Mediterranean 1942;

ADM 1/14309: Awards to Fleet Air Arm Officers of HMS FORMIDABLE and 821 Squadron at Malta 1943–1944;

ADM 1/14359: No. 841 Fleet Air Arm Squadron: operational awards to personnel 1943;

ADM 1/29930: New Years Honours List 1945: Fleet Air Arm Personnel;

ADM 1/30325: Awards to 9 officers and men of 819 Squadron Fleet Air Arm for services against enemy light forces operating from Channel bases.

ADM 116 contains a number of administrative files relating to the FAA, including ADM 116/5468: Operation Tungsten: report of Fleet Air Arm attack on German battleship TIRPITZ: recommendations for honours and awards. ADM 223/336 is a Naval Intelligence report on the Taranto attack of 1940 and even the War Office chips in with a report WO 203/220: Fleet Air Arm: operations, South East Asia, April 1944–April 1945.

The part that Fleet Air Arm aircraft took in particular actions will be recorded in the general reports of those battles so it is always worth searching for these. HMS *Formidable* took part in the Battle of Matapan in March 1941 when her aircraft damaged an Italian battleship and cruiser. There are reports on the battle in ADM 199/781: Battle of Cape Matapan, and ADM 199/1048 and 1049: Naval operations in the Mediterranean including the Battle of Cape Matapan. ADM 1/11448 contains Awards to personnel of HMS FORMIDABLE for courage and good services in Battle of Cape Matapan.

HMS *Formidable* also took part in operations against the German battleship *Tirpitz* in 1944 and one such attack was codenamed 'Mascot'. A search under Mascot brings up ADM 1/29792: Awards to 2 pilots of HMS Formidable for services in attacking a German destroyer in Norwegian Waters during an air strike on the battleship Tirpitz July 18 1944 (Operation MASCOT). A search under *Tirpitz* also reveals ADM 1/16695 Awards to personnel of HM Aircraft Carriers FORMIDABLE, FURIOUS and INDEFATIGABLE for air attacks on German battleship TIRPITZ, 22–29 August 1944.

Even though the FAA were a naval responsibility there are a number of records in the AIR class including: AIR 23/6377: AHQ British Air Forces Greece, Operation carried out in conjunction with Navy and Fleet Air Arm; AIR 23/6792: RAF Middle East Operations by Fleet Air Arm Units 1943; AIR 20/1239: Fleet Air Arm squadrons for operation 'Overlord'. These deal with operations, and there are numerous files on aircraft and personnel requirements.

Combat reports for Fleet Air Arm squadrons 800, 803, 804, 808, 820, 825, 841, 885, 886, 887, 894 and 897 are in AIR 50 series between AIR 50/320 and AIR 50/331.

Ship's logs, which are mainly concerned with the ship's movements and weather conditions but which will give some details of operations, are in ADM 53. HMS *Ark Royal*'s logs for September 1939 to September 1941 are between ADM 53/107522 and ADM 53/113633; those for HMS *Formidable* are between ADM 53/112247 and ADM 53/1121385 (November 1940 and August 1945).

Fleet Air Arm service records

Between 1945 and 1973 every sailor was given his service record (in the form of a discharge certificate) when he left the navy. This means that if you have the discharge certificate you don't need to apply for a service record. If you don't have it then records for members of the Fleet Air Arm are still held by the Ministry of Defence, and you will need to contact: NPP Accounts, 1, AFPAA, Centurion Building, Grange Road, Gosport, Hampshire, PO13 9XA.

You will need to be able to prove that you are next of kin, or provide their authority (or the authority of the serviceman himself), to receive a copy of the record. The RN Record Office will provide you with a Next of Kin form to complete which will explain who the next of kin are (all children count equally for example). It may occasionally be necessary to provide proof of identity. There is a search fee (2007) of £30.

Army pilots in the Second World War: air observation posts

One of the great successes of the First World War had been the splendid co-operation between aircraft and artillery but, due to inevitable budget cuts, between the wars the army had to rely on just a few Army Co-operation Squadrons. Dissatisfied with the delays that inter-service liaison inevitably caused, a small group of army officers began secret experiments in the late 1930s using light aircraft and voice radio to spot for their artillery. Their success led to the setting up of the first Air Observation Post (OP) squadrons in which RAF ground crew serviced the aircraft but the pilots were all artillery officers. Flying Auster aircraft from temporary air strips close to the front line, these aircraft supplied vital information for the artillery on all fronts from 1942 onwards. Within forty-eight hours of D-Day two Air Ops squadrons were flying off French soil in support of the guns. Shortly before the end of the Second World War the first helicopters were used experimentally directing artillery fire.

The squadrons were allocated numbers between 651 and 660. Their Operation Record Books are in AIR 27 but the information they provide can vary depending upon which officer wrote up the record. 653 Squadron was formed on 20 June 1942 at Old Sarum and went to Normandy on 27 June 1944. It was disbanded in Germany on 15 September 1945; its ORB is in AIR 27/2171.

The ORB for June and July 1944 contains a general survey of events which describes the squadron's activities in the Normandy bridgehead as part of 12 Corps.

> The large majority of the early work in July was anti-mortar sorties designed firstly to spot the mortars firing (particularly nebelwerfers) and secondly to

> stop the firing. This latter was successful up to a point – the Bosche thought at first that the Air O.P. could see the mortars firing but later discovered that we could not.

On 17 July the squadron had its first casualty when Captain Watkinson was hit by a British shell while returning to his Advanced Landing Ground (ALG). Another danger of being based so close to the guns was demonstrated when 'B' Flight's ALG was bombed at night by German aircraft attacking the nearby battery. 'All four aircraft were written off, all the tentage, one pilot received a bullet through the foot and was evacuated to the UK.' During the last part of July the squadron experimented with aerial photography with great success. During that month they flew 610 sorties and had one flying casualty.

Given their dual nature it is perhaps unsurprising that records relating to AOPs are split almost 50/50 between the AIR and WO series. Not only are there ORBs in AIR 27 (652 Squadron's is in AIR 27/2170 covering May 1942 to April 1946) but also some War Diaries in the records of the various theatres that the squadrons fought in. 652 Squadron was with the North West Europe Expeditionary Force and its diary for the period January to December 1945 is in WO 171/4749. Also in the WO 171 series are War Diaries for 653, 658, 659, 660 and 661 Squadrons. 659 Squadron, which also fought in Burma, has its War Diary in WO 172/7396. No War Diaries appear to have survived for AOPs based in Italy but there is one record in WO 204/7535 for 654 Squadron that appears to relate to training.

There are a number of recommendations for gallantry awards for Army personnel in the Air Observation Posts in WO 373, which can be searched using the proposed recipient's surname.

The Glider Pilot Regiment

In 1940 the Germans showed the value of the military glider when they were able to land, with great precision, specially trained attack and demolition units right on top of the 'impregnable' Belgian fortress of Eben Emael which fell shortly after. Churchill himself demanded the establishment of equivalent British units. There was much scorn poured on the initial idea of the Glider Pilot Regiment.

> The idea that semi-skilled, unpicked personnel, (infantry corporals have I believe even been suggested) could with a minimum of training be entrusted with the piloting of these troop carriers is fantastic.

So wrote the Deputy Chief of the Air Staff. The Glider Pilot Regiment (GPR) was created, however, and went on to prove its worth. The GPR pilots were trained not only to fly and land their gliders, with their cargo of troops or equipment, but to fight alongside the troops they carried. The GPR carried troops into action in Sicily in 1943, at D-Day, in the Balkans, southern France and Arnhem in 1944, in India and Burma 1944/5 and at the Rhine crossing in 1945.

Operation Record Books of the Glider Training units and schools are in AIR 29. Among these are:

AIR 29/524: Glider Training Squadron, Haddenham, later No. 1 Glider Training School, Thame, Croughton 1941 Jan.–1946 May

AIR 29/526: Glider Instructors School, Thame, 1942 Oct.–1943 Jan.

AIR 29/819: No. 2 Heavy Glider Maintenance Unit, Snailwell, 1943 Aug.–1944 March.

As Army Units the Glider Pilot Regiment kept War Diaries, much as the RFC had done at the beginning of the First World War. These are arranged by army operational theatre: WO 166 covers Home Forces, WO 169 covers North Africa, WO 170 Italy and WO 171 North West Europe. This means that the sequence of diaries for a given unit may have to be looked for across more than one TNA Class. The 1st Battalion of the Regiment was formed in the UK so its early War Diaries are in WO 166 – it later went to North Africa and to North West Europe forces so that the run of War Diaries is spread:

WO 166/10470: 1 Bn. Glider Pilot Regt. War Diary, 1943 Jan.–May

WO 169/10341: 1 Bn. Glider Pilot Regt. War Diary, 1943 July–Dec.

WO 171/1233: 1 Glider Pilot Regt. 1944 Jan.

There are a number of recommendations for gallantry awards for Army personnel in the Glider Pilot Regiment in WO 373, which can be searched using the proposed recipient's surname.

The Museum of Army Flying near Andover holds much material, including the memoir of Glider Pilot Geoffrey Barkway, who crash-landed his glider close to the Orne Bridge of D-Day, carrying troops who seized the vital crossing. Many years later he was unstinting in his praise of both the RAF pilots who got them to Normandy and to his fellow glider pilots.

> However good Major Howard and his lads were, if we hadn't dropped them there, then the gaffe would have been blown and confusion would have reigned. Not that it didn't reign anyway, but it reigned in the right place.

Service records for Army personnel

Records for members of the Army Air Corps and Glider Pilot Regiment are still held by the Ministry of Defence, so to obtain copies you'll need to apply to: The Army Personnel Centre, Historic Disclosures, Mailpoint 400, Kentigern House, 65 Brown Street, Glasgow, G2 8EX.

You will need to be able to prove that you are next of kin, or provide their authority, or the authority of the serviceman himself, to receive a copy. The Army Personnel Centre will provide you with a Next of Kin form to complete which will explain who the next of kin are (all children count equally for example). It may occasionally be necessary to provide proof of identity. There is a search fee (2007) of £30.

The Air Transport Auxiliary

The Air Transport Auxiliary, which was formed in 1939, was a civilian organization whose members were employed by BOAC and which delivered aircraft from the factories to the RAF stations. Though the vast majority of its pilots were men it did employ a small number of female pilots (including Amy Johnson), as well as an administrative staff. All of their personal records of service (which would otherwise have been destroyed) are preserved by the RAF Museum Hendon archive, though a seventy-five-year embargo exists if you are not next of kin. Records include those of administrative, catering and other staff as well as of the pilots.

You will need to be able to prove that you are next of kin, or provide their authority (or the authority of the person themselves) to receive a copy of the record.

Chapter 8

THE RAF AFTER 1945

Post-war RAF records

At the end of the Second World War the RAF had some 460 squadrons, over 9,000 aircraft, 1,200 airfields and over 1,000,000 service personnel. The first priority was to demobilize the men and women who had been called up for war service and the winding up of the various units can be followed through their ORBs. There are various policy files relating to Demob in the AIR 2 series, notably AIR 2/5292: Demobilization policy; AIR 23/1163 and AIR 23/1164: Post-war release and demobilization, organization policy; and AIR 23/2321 and AIR 23/2322: Demobilization planning and release scheme: policy. By 1947 the RAF had reduced to below 300,000 personnel.

Fortunately for the family historian the record types remain the same as during the war: AIR 24, AIR 25 and AIR 26 still cover commands, groups and wings and AIR 27, AIR 28 and AIR 29 squadrons, stations and miscellaneous units respectively. Some of the various bases used specifically in demobilization have their ORBs in AIR 29, specifically: AIR 29/930: No. 1 Demobilization Centre, RAF Walton, later No. 1 Demobilization Centre, Royal Indian Air Force, Walton; AIR 29/1646: Overseas Release Centre, Singapore; AIR 29/505: No. 100 Personnel Dispersal Centre, Uxbridge; AIR 29/1097: No. 2 Overseas Release Centre, Fort de L'eau, later No. 2 Release Embarkation Centre CMF.

The RAF was important in the post-war withdrawal from Empire and for the rapid movement of troops. It was certainly felt at the time that RAF personnel were being deliberately kept on longer than Army and Navy servicemen and this led to a series of 'strikes' particularly in the Far East. There is coverage of these in various files, in particular:

AIR 20/9245: Courts of Inquiry and Inquests: Strikes in South East Asia Command;

AIR 23/1986: Brief history of events associated with disaffection and 'strikes' among personnel in RAF units;

AIR 23/2315: Discipline: Air Ministry Court of Inquiry into strikes;

AIR 20/9244: Incidents at RAF stations in South East Asia Command.

Throughout the 1950s and early 1960s Britain continued to attempt to act as a major world power, with commitments in the Middle and Far East, Europe and

Africa. The start of the Cold War in 1948 also meant that a new enemy, the Soviet Union, had to be taken into account. Bases were kept in Germany and the RAF also maintained Britain's nuclear deterrent. There are numerous files both on deterrent policy and nuclear weapons generally, including:

AIR 2/13707 to AIR 2/13715: Maintenance of the British nuclear deterrent 1953–1965;

AIR 8/2311: British controlled contribution to the nuclear deterrent;

AIR 19/940: V bomber force: deterrent;

AIR 20/12512: Procedure for launching nuclear deterrent;

AIR 2/13684 to AIR 2/13688: RAF policy requirements for nuclear weapon trials;

AIR 2/13729 Nuclear weapons effects, tactics and targets;

AIR 2/18095 to AIR 2/18102: Joint Planning Staff Working Study: Global Nuclear War in 1957;

AIR 20/10115: Command organization in nuclear war.

Army aviation records after the Second World War

After the Second World War the Air Observation Posts and Glider Pilot Regiment (now flying light aircraft) served in Malaya, Korea, Eritrea, Cyprus and Suez. In 1957 the two merged to create the Army Air Corps (AAC) and absorbed the Joint Experimental Helicopter Unit (JEHU) which had been carrying out the army's first experiments with helicopters. Since 1957 the AAC has been increasingly flying light reconnaissance helicopters providing support for observation and reconnaissance, artillery fire control, limited movement of men and materials and liaison.

There are a few ORBs for Air Observation Posts for the early 1950s in AIR 29 series. War Diaries for the Army Air Corps since 1957 are in TNA's WO 295 series. The ORBs for No. 2 Reconnaissance/Liaison Flight between 1957 and 1962 are in WO 295/1, WO 295/2, WO 295/14 and WO 295/15, for example, and the War Diaries for 20 Independent Reconnaissance Flight for January to March 1958 are between WO 295/44 and WO 295/46.

The Fleet Air Arm after the Second World War

Fleet Air Arm squadron diaries for the late 1940s are in TNA series ADM 207 following on from ADM 207/54. With the exception of ADM 207/68, which has the diaries for 1832, 1853 and 1836 Squadrons between September 1954 and September 1957, they run until the end of 1950.

Squadron Record Books after 1950 are held by the Fleet Air Arm Museum at Yeovilton (see Appendix 9), which also has a few from the 1940s. In addition the museum holds the Squadron 'Line Books' which are particularly amusing. Running in parallel with the Squadron Record Books these are a much less formal view of the squadron's activities. Concentrating mainly on the aircrew, they consist of informal (sometimes extremely informal!) photographs, newspaper cuttings, cartoons and poems. For most of them you get the feeling that you need to have been there to understand the exact nature of the joke, but they do give a feel for the life of an FAA crew member at work and at play. Most of the Line Books cover peacetime operations but there are one or two from the Second World War and others from Korea and other wars.

Under the thirty-year rule the archive will not allow access to Record Books or Line Books less than thirty years old except to service personnel.

National Service

National Service was introduced at the beginning of the Second World War but was always intended to cease at the end of hostilities. In 1948 a National Service Act was passed, effective from 1 January 1949, which made all young men liable to service of eighteen months with four years in the Reserve; then changed to two years' service with 3½ years in the Reserve. For a technical service such as the RAF this barely gave time for a man to be properly trained before he was discharged. The Reserves were called up during the Suez Crisis to replace men sent out to the Middle East. It was promised early on that no one born during Second World War would have to do National Service so it was inevitable that it would cease by about 1963.

A National Serviceman's service record

My father, Philip Valentine Tomaselli, was called up for his National Service in the RAF on 12 February 1959. He kindly applied for a copy of his service record for me and a covering letter from the Royal Air Force Personnel Management Agency explains that, in accordance with Ministry of Defence document retention policy, most of his documents had been destroyed. His RAF Form 543 Record of Service was preserved and a copy is reproduced on page 130.

It shows his next of kin (my mother), details of their marriage, civilian occupation, date of birth, his service number (5066923), his various postings, promotions and qualifications and conduct and trade assessments. There is space for medals, honours and awards, for prior service in HM Forces and for time forfeited due to misconduct (none of which applied to him). He was first sent to No. 2 Reception Unit at RAF Cardington to be kitted out, then for basic square bashing training at No. 4 School of Recruit Training at RAF Wilmslow. Following this he was trained at No. 2 Radio Training School at RAF Yatesbury to be a radar fitter for the V-bomber force (the Vulcan and Victor aircraft that carried Britain's nuclear weapons). From here he did his actual service at RAF Lindholme (Bomber Command Bombing School) where he worked on ground training equipment until he was discharged to the reserve in March 1961.

The records for his various stations can be traced at TNA in the same manner as for other periods after the First World War. AIR 29/2991 and AIR 29/2717 are the ORBs for No. 2 Reception Unit, Cardington, covering December 1955–February 1961; AIR 29/2640 is the ORB for No. 4 School of Recruit Training, Wilmslow, for January 1956–November 1959. AIR 29/2336 covers RAF Yatesbury (No. 2 Radio School) January 1956 to December 1959. Unless your relation was an officer it is highly unlikely that they'll be mentioned.

The ORB for Yatesbury for September 1959 mentions that the weather was warm and dry with long periods of sunshine. The station was visited by Commandant Karam of the Lebanese Air Force to check the signals training being given to some

5066923 | TOMASELLI | Philip Valentine
NUMBER | SURNAME | CHRISTIAN NAMES

RECORD OF SERVICE (AIRMEN)

R.A.F. FORM 543R (Revised March 1955)

NEXT OF KIN	MARRIAGE STATE M
DATE 10 - 2 - 59	TO ARMSTRONG, Doreen (British Spinster)
NAME MRS D TOMASELLI	PLACE St. Margaret's Church DATE 29.3.58
ADDRESS 33 CHIRTON GREEN. NORTH SHIELDS. NORTHUMBERLAND	Scotswood,
RELATIONSHIP WIFE	P.O.R. No. Newcastle-upon-Tyne, Northumberland
OTHER PERSON TO BE NOTIFIED	MISCELLANEOUS
NAME	N.H.S. NUMBER GBBF.72.3
ADDRESS	
RELATIONSHIP	'G' RESERVE LIABILITY
DISCHARGE DETAILS	TERMINATED ON 30 JUN 69
DATE AND AUTHORITY	
CAUSE	
TOTAL SERVICE TO COUNT YEARS DAYS	MOD LETTER AF/CT25/66M11 (RAF)
TOTAL QUAL. SERVICE YEARS DAYS	DATED 26 FEB 69 REFERS
PENSION AWARD	
ADDRESS ON DISCHARGE 20/61 (9621)	
33 CHIRTON GREEN, NORTH SHIELDS, NORTHUMBERLAND	
MEDALS, HONOURS AND AWARDS, ETC.	
	OVERSEAS SERVICE
	SCORE / DATE — VETTED — SCORE / DATE — VOL. — SCORE / DATE — P.W.R.

5066923 | TOMASELLI | Philip Valentine
NUMBER | SURNAME | CHRISTIAN NAMES

MOVEMENTS

DATE	UNIT TO	AUTHORITY	CODE
12 Feb.59	2 R.U.		
20.2.59	4 S of RT WILMSLOW	4SRT 7/59	
29.4.59	2RS YATESBURY	YSRT/PO/9/59	
8 JAN 60	B.C.B.S. LINDHOLME	7/2A/20026/60	
11 FEB 61	DISCH WTS QR 607/5	20/61	
12 MAR 61	ENLIST H RES	(9621)	
7-MAR-61	DISCH WTS QR 607/5	22/61	
8-MAR-61	ENLISTED H RES		9621

MUSTERING, REMUSTERING AND TRANSFERS

AUTHORITY	TRADE	EFFECT. DATE
	Recruit	12 Feb.59
YSRT/PO/9/59	RAD ASST / AIR RADAR FITTER	29.4.59
658/59 (0565)	E MECH G 11 AIR RADAR FITT	12-2-59
621/60 (0565)	AIR RADAR FITT (002)	6-JAN-60

PROMOTIONS, REVERSIONS

AUTHORITY	DESCRIPTION	EFFECT. DATE
	A.C.2	12 Feb.59
608/59 (0565)	SAC	12-2-59
621/60 (0565)	J/T	6-JAN-60

H87932 Wt.64900-BK.2739 105M(9) 5/55 Gp.840 F. & C. Ltd., London

A National Serviceman's service record – excerpt from the author's father's service record, showing how little they've changed over the years. (Mr P V Tomaselli)

of his officers and a new Station Commander, Group Captain C H Press, was appointed at the end of the month. As at 30 September the strength of the station was 106 officers and 612 other ranks on the permanent staff, with five officers and 1,690 other ranks in training. There were fourteen foreign personnel in training, including one Iraqi, one Iranian, three Burmese and two Lebanese.

Operations Records Books for the Bomber Command Bombing School at RAF

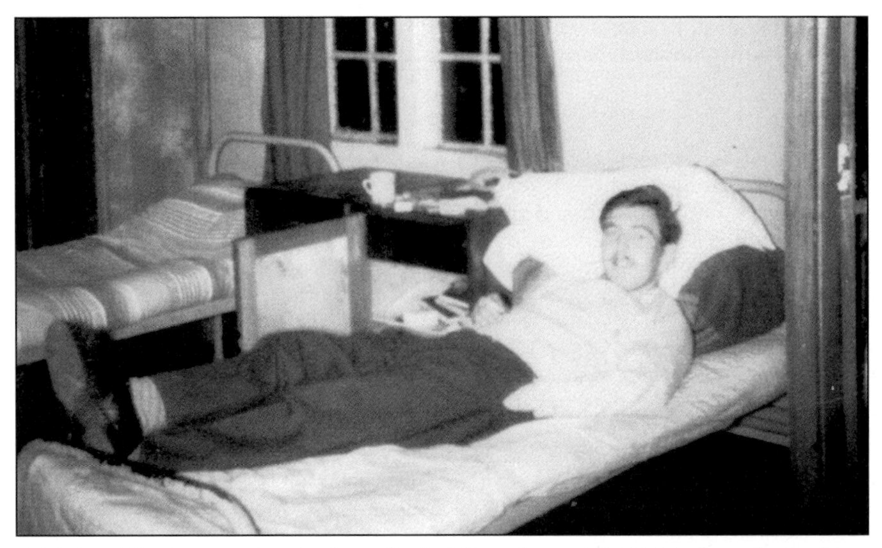

A National Serviceman at rest – the author's father relaxes at RAF Yatesbury during his service there in 1959. (Mr P V Tomaselli)

Lindholme, covering January 1956 to May 1961, is in AIR 29/2693. One of my father's abiding memories of Lindholme is of virtually the entire station, including all the officers in their best mess kit, being called out to fight a serious fire on the moors. A note in the ORB for April 1960 confirms:

> A Dining-in night was held at the Officer's Mess on 21st April, this was interrupted however by the Station Commander's announcement that a fire on the moors was getting out of control. Those dining-in together with a large number of airmen from the station went to the moors to fight the fires. A message of appreciation was subsequently received from the Chief Fire Officer.

The ORB for RAF Yatesbury recorded:

> The last National Serviceman at Yatesbury was discharged on 1st November 1962. He was 5082101 Cpl. Technician J M J Clarke an education instructor. He reported to Yatesbury on 1st November 1960 on completion of basic training and served throughout at this station.

RAF operations from the Second World War to 1960

In addition to the usual unit and formation ORBs, which continue post-1945 in the same series as previously, there are numerous records in the form of policy documents, reports on operations and organization which might yield useful information to the diligent researcher.

There are general histories of some of the campaigns, compiled and written by the Air Historical Branch in the AIR 41 series which give useful lists of squadrons and other units that took part in operations.

Palestine

Several RAF Squadrons and support units were based in Palestine during the increasingly serious disturbances that led up to the British evacuation in 1948, the partition that followed, and the first Arab–Israeli war. They provided aerial reconnaissance for anti-terrorist operations and conducted patrols intercepting shipping carrying illegal immigrants. Several aircraft were lost on the ground to terrorist bombs. When the final withdrawal was in progress the Arab–Israeli war had already broken out and, as both sides used British aircraft, there were occasional incidents where the RAF was attacked by both sides, each mistaking them for the enemy, and some casualties occurred. Many of the Squadron ORBs for the period seem to be missing, though sometimes their appendices, containing copies of orders and reports, are extant. This can mean a bit of a search looking for station, wing or group ORBs in order to fill in some of the fine detail.

AIR 23/8350 contains an invaluable report 'Evacuation of R.A.F. from Palestine', by AVM W L Dawson, AOC Levant, which gives a lot of the background and detail to the campaign. Other reports which might be of use include:

AIR 20/4964: Situation in the Levant, May 1945 to March 1946;

AIR 20/6377 to AIR 20/6380: Intelligence Section AHQ Levant: monthly air letters; operational and intelligence reports, 1946 to 1948;

AIR 20/10136 and AIR 20/10137: Palestine: Arab/Israeli affairs;

AIR 20/8889 and AIR 20/8890: Defence and evacuation of Palestine;

AIR 23/8331: Withdrawal of British forces from Palestine.

The Berlin Airlift

In 1948, in an attempt to persuade the British, French and Americans to leave Berlin, which was isolated within the Soviet Zone of Occupation in Germany, the Russians imposed a land blockade in the hope of starving them out of the city. It was clear that any attempt to break the blockade by land would result in war but, fortunately, three air corridors into the city from the west had already been agreed. It was estimated that 1,500 tons of foodstuffs alone would be required per day to keep the citizens from starvation, along with more than 1,500 tons of coal for heating and power. In a herculean effort the United States Air Force and Royal Air Force flew the required amounts in, using the two existing airfields at Templehof and Gatow, a new field built at Tegel, and even flew supplies in to Lake Havel in RAF Sunderland flying boats. In May 1949 the Russians lifted the blockade. The Americans had lifted nearly 1,800,000 tons and the RAF nearly 542,000.

The British side of the operation was known as 'Plainfare'. AIR 10/5067 contains the comprehensive 'A Report on Operation Plainfare: (The Berlin Airlift) 25.6.48–6.10.49' and there is another in AIR 10/8955. This is another case where it is useful, when searching on TNA catalogue, to look not only for 'Plainfare' but also for 'Berlin' as, though there is substantial overlap, a number of important records come up under one but not the other. Among the records that come up purely under 'airlift' are:

AIR 2/9986: Awards for Berlin airlift;

AIR 20/7282: Berlin Airlift 1948/9: daily airlift reports (ALREPS), nos. 15–98;

AIR 20/7809: Berlin Airlift: HQ Airlift Intelligence Staff: situation reports;

AIR 38/377: Report on activities of Combined Airlift Task Force: Berlin Airlift.

Amongst the records that come up purely under 'Plainfare' are:

AIR 20/6891 to AIR 20/6894: Germany (Code 55/2/4): Operation PLAINFARE;

AIR 38/309: Operation 'Plainfare' : employment of glider pilots;

AIR 55/117: Interim report on operation 'Plainfare': RAF Station, Lubeck.

It is worth pointing out that, in addition to the AIR files, there are forty files specific to the airlift in the Foreign Office (FO series). Depending on how detailed you would like your research to go, it can be useful looking about for associated records. RAF operations didn't exist in a vacuum and there could be interesting background information in War Office (Army), Naval (ADM), Foreign Office (FO) or Colonial Office (CO) files.

Korea

The involvement of the RAF in the Korean War (1950 to 1953) was small in terms of numbers of units employed, though several aircrew served with both the United States Air Force and the Royal Australian Air Force. Three squadrons of Sunderland flying boats (nos. 88, 205 and 208) flew maritime reconnaissance missions from Japan and Singapore. There was further involvement from the Fleet Air Arm and the Army's Air Observation Posts (see the sections on the Fleet Air Arm and Army for more details).

Considering the small part played by the RAF in the war there are a surprising number of records in the AIR series (seventy-nine in all). Some of them that may be of interest include:

AIR 8/1782 to 1784: Korea: air action;

AIR 8/2473: Treatment of returned POWs from Korea;

AIR 8/1709: Korea – MIG15/Meteor combats: progress and tactical, narrative reports; story of F86 v MIG15 by Squadron Leader W Harbison;

AIR 20/7611: Korea: MIG15/Meteor combats;

AIR 20/7497: Korea: casualty air evacuation;

AIR 20/11380 POWs: policy, instructions to servicemen on their conduct when captured and details of British POWs' experiences in Korea;

AIR 20/7412 Sunderland operations in Korea and maritime warfare policy;

AIR 2/12276: Awards by the USA to RAF and Dominion Air Forces in Korea;

AIR 40/2642 Korea: evasion and escape reports.

Malaya

During the Second World War the British provided communist guerrillas in Malaya with large quantities of guns and ammunition to fight the Japanese. In 1948 these were turned against the British administration and a ferocious guerrilla war began that lasted until 1960. The RAF provided transport and supplies for troops operating in deep jungle, as well as flying strike operations against the insurgents. Malaya saw the first use of RAF helicopters for casualty evacuation when the Casualty Evacuation Flight was established in 1950.

AIR 41/83: 'The Malayan Emergency, 1948–1960' is the campaign history. Other records (among very many) which may be of interest are:

AIR 22/525 to AIR 22/527: Air Headquarters Malaya weekly summary of air operations, January 1952 to April 1953;

AIR 8/1924 Malaya: internal security 1956–1957;

AIR 20/9271 Malaya: air support for operations by Lincoln aircraft 1953–1954;

AIR 20/6656: Use of helicopters for casualty evacuation in Malaya;

AIR 20/9448 to AIR 20/9452: Malaya: internal security 1953 to 1959;

AIR 23/8592: Chemical attacks on Communist terrorist cultivated areas in Malaya;

AIR 23/8618: RAF Regiment (Malaya): operations reports.

As the RAF worked so closely with the army and police during the emergency this is another case where searches amongst files in the WO and Colonial Office records are likely to produce useful information.

Kenya

The RAF played only a small role in operations in Kenya during the Mau Mau rebellion of 1952 to 1956, dropping leaflets, evacuating casualties and supporting ground troops. 50 and 66 Squadrons of the RAF Regiment were also involved on detached duty from Aden.

AIR 8/1886: Air operations in Kenya;

AIR 20/9041: HQ Middle East Air Force: air operations in Kenya against Mau Mau;

AIR 20/9515 to AIR 20/9517: Kenya emergency: summaries of terrorist and security forces action;

AIR 20/9530: RAF operations in Kenya;

AIR 14/4071 to AIR 14/4073: Operations in Kenya: intelligence reports;

AIR 14/4496: Nos. 49/102 (B) squadrons' role in Mau Mau operations Kenya;

AIR 23/8615: RAF Nanyuki, Kenya: Mau Mau operations.

This is another case where, as the RAF worked so closely with the army and police during the emergency, searches amongst files in the WO and CO records are likely to produce useful information.

Suez

The Suez Crisis of 1956 is generally remembered for the landing of paratroops and Royal Marines to seize the Canal, but for nearly a week before the landings the RAF and Fleet Air Arm flew a series of missions to destroy the Egyptian Air Force and to reduce the Egyptian Army. It was the biggest RAF active service operation in the period. It was the RAF's first war fought mainly with jet aircraft in both fighter and bomber roles, and saw the very first use of helicopters to land troops from the sea.

Operations commenced on the night of 31 October as aircraft from Nicosia and Malta, many of which had been moved to the Mediterranean only a few days before, attacked Almaza and Kabrit airfields. In the early hours of the next morning further raids were carried out on airfields at Abu Sueir, Cairo West and Luxor and on 2 November another raid on Luxor targeted the Egyptian aircraft that had survived the previous attacks. On 3 November the second stage of operations saw eight squadron aircraft attack the barracks at Almaza in two waves. Next day, in their final raids of the campaign, six crews attacked Huckstep Ordnance Depot and scored several hits on a large tank and ammunition depot

Fleet Air Arm squadrons 830, 892, 893, 897, 899 and 849A Flight left Malta on 29 October on HMS *Eagle* and was in position off the Egyptian coast by 31 October. *Eagle*'s first strike attacked Inhas airfield, which was still being attacked by RAF Canberras when her Seahawks and Venoms commenced their raid. Further strikes during the day hit Dikhelia, Cairo West and Abu Suier airfields.

After five days of aerial attacks the landing of troops by parachute, assault craft and, for the first time, helicopter took place on 6 November. All operations were judged a great military success but the operation was politically grossly misjudged and a cease fire was soon imposed by the United Nations, when the USA refused to back Britain and France.

The RAF and FAA flew thousands of sorties during the war. RAF Canberras flew 72 missions from Malta, and Valiants an additional 49, dropping at least 1,439 bombs. HMS *Eagle*'s fighters flew 621 combat sorties, and 415 were flown off from HMS *Albion*, dropping 229 bombs and firing 1,448 rockets and 88,000 cannon rounds. One Canberra, one Venom, two Sea Hawks, two Wyverns, and two Whirlwind helicopters were shot down during the fighting, and over fifty other aircraft were damaged.

Many files on the planning of air operations can be found in the AIR 20 series. It is worth searching not only under 'Suez' but also under 'Musketeer' which was the code name for the whole operation. The Diary for the Joint Helicopter Unit, which took part in the air landings and which was soon to amalgamate into the Army Air Corps, is in WO 288/136. AIR 20/10210 discusses 48 Field Squadron, RAF Regiment's role and deployment in Suez. ADM 116/6134 contains HMS *Eagle*'s report of proceedings including descriptions of the first air landing of Royal Marines.

Building a bigger picture: other files relating to the Suez Crisis

The enormous scale of the Suez operation and its political ramifications, which spread around the globe, mean that there are scores of files available in the AIR series alone. Some are:

AIR 8/1939: Suez Canal crisis: photographic reconnaissance

AIR 8/1948: Suez Canal crisis planning for military operation MUSKETEER

AIR 8/1951: Operation MUSKETEER: situation following ceasefire

AIR 8/2071: Malta: implications of Suez crisis

AIR 8/2077: Jordan: implications of Suez crisis

AIR 8/2085: Suez Canal crisis: evacuation of British subjects from Middle East countries

AIR 8/2086: Suez Canal crisis: press security

AIR 8/2089: Suez Canal crisis: personnel questions

AIR 8/2090: Suez Canal crisis: RAF deployment

AIR 8/2093: Suez Canal crisis: Operation GOLDFLAKE

AIR 8/2097: Operation MUSKETEER: sitreps (situation reports)

AIR 8/2099: Operation MUSKETEER: reports on damage and casualties

AIR 8/2111: MUSKETEER: air operations

AIR 8/2125: Suez Canal dispute (Operation MUSKETEER): honours and awards

AIR 14/4030 and AIR 14/4031: Operation MUSKETEER: analyses of operations

AIR 14/4441: Bombing and ground attack operations during Operation MUSKETEER

AIR 19/857: Suez Canal operations

AIR 20/9225: Suez Canal crisis: intelligence on Egypt

AIR 20/9551: OPERATIONS: North and East Africa (Code 55/2/8): Operation MUSKETEER Air Task Force HQ Files: Medical and casualty air evacuation: correspondence and signals

AIR 20/9553: Operation MUSKETEER Air Task Force HQ Files: Employment of ground attack aircraft 'D' and 'D+1'

AIR 20/9560: Operation MUSKETEER Air Task Force HQ Files: Operation ALACRITY

AIR 20/9678: Operation MUSKETEER: Suez Canal salvage operations

AIR 20/10129: Suez Canal crisis: 249 Squadron, Amman

AIR 20/10203 and AIR 20/10204: Operation MUSKETEER: bomber operations

AIR 20/10210: Operation MUSKETEER: No. 48 Field Squadron, RAF Regiment; role and deployment

AIR 20/10211: Operation MUSKETEER: Allied Air Forces; order of battle

AIR 20/10214: Operation MUSKETEER: air operations

AIR 20/10215: Operation MUSKETEER: targets

AIR 20/10352: Operation CHALLENGER/MUSKETEER: moves of RAF Regiment Squadrons and reports

AIR 20/10363: Operation MUSKETEER: HQ Air Task Force operational reports

AIR 20/10369: Operation MUSKETEER: psychological warfare

In addition there are some 2,696 Foreign Office (FO) files, 18 Cabinet Office (CAB), 58 Prime Minister's Office (PREM), over 80 War Office (WO) and some 80 Admiralty (ADM) files. Some of these were only released at the end of 2006.

Cyprus

Cyprus provided some of the main bases for the Suez operations but was, at the same time, the focus for EOKA terrorist operations trying to secure independence from Britain. The RAF, reinforced by Army helicopter units and the Fleet Air Arm, flew anti-terrorist patrols, transported troops and made occasional air strikes on terrorist bases. Cyprus was granted its independence in 1960 but continues to lease air bases to the RAF.

Because of Cyprus's perceived strategic value for the RAF in allowing them to cover the whole of the Eastern Mediterranean you will find a lot of records dealing with the granting of independence and the retention of the bases, as well as with the island's role in the Suez crisis.

Records in the AIR series relating specifically to the terrorist situation are:

AIR 8/1921 to AIR 8/1923: Cyprus: internal security, 1956–7;

AIR 20/10330: Cyprus: internal security, 1953 to 1963;

AIR 20/9280: Provision of helicopters for internal security role in Cyprus;

AIR 20/8894: Attacks on personnel and in Cyprus by terrorist organisation EOKA.

In addition to the operations detailed above the RAF operated in support of ground troops or carried out transport operations almost all round the globe and search of TNA catalogue using the place name, country or area of the world (i.e. South East Asia) should locate documents of interest.

An excellent website http://britains-smallwars.com is devoted to Britain's post-1945 conflicts. As well as providing invaluable information on the units employed and the work they did, it frequently contains memoirs of participants and photographs of men and aircraft involved.

Freedom of Information Act

On occasion you may come across a file that is at the National Archive but which is marked 'Closed Or Retained Document' but which you think may be of interest to you. One such is AIR 40/2629: 'Korea: British PoWs; collaborators and security suspects'. Under the Freedom of Information Act it is possible to request a review of some of these files with a view to having them opened, using TNA's catalogue. There is no need to provide a reason, and you will be advised of the decision in due course.

The RAF after 1960

The scope of this book ends in the early 1960s, with the end of National Service, but the RAF continued to serve throughout the world on active service (Aden, Borneo, Northern Ireland, the Falklands, the Gulf War, Bosnia, Kosovo, Sierra Leone, Afghanistan and Iraq). Records of their involvement continue to be regularly released to the National Archive under the thirty-year rule. The latest releases date from 1976 at time of writing, so that the ORB for No. 11 Maintenance Unit RAF Chilmark up to 31 December 1976 is now available in AIR 29/4514. Some sensitive documents, for example, AIR 20/12877: 'Study of reconnaissance in support of security forces in Northern Ireland', do remain closed for security reasons, others may be requested for review under the Freedom of Information Act.

It is over 100 years since the first Royal Engineers took to the skies in their balloons, and nearly 100 years since Colonel Cody first flew for them in a heavier than air machine in 1908. In that time its officers and serving men and women have flown and fought in almost every corner of the globe. If your relative was among them you should be honoured. Enjoy your research.

Chapter 9

MEDALS, CASUALTIES AND COURTS MARTIAL

The whole subject of medals, honours and awards is one deserving of a book in its own right, so this section is designed only as a basic guide of where you are likely to find the simplest records. For a more detailed approach readers are recommended to consult, William Spencer, *Medals: The Researcher's Guide* (National Archives Publications, 2006).

There are basically three types of medals that can be awarded: campaign medals, awarded for taking part in a particular war or campaign, awards for gallantry or meritorious service and long service and good conduct medals.

Campaign medals before the First World War

It is always possible that someone serving in the early days of the RFC or the RNAS might have been awarded medals for prior campaigns, in particular the Sudan (1896–9), the Boer War (1899–1902), the Boxer Rebellion (1900), Somaliland (1900 onwards) or any of the small frontier wars in India. These medals would have been for their service before they joined the air services and should be borne in mind if you're looking at the career of anyone who saw much service before transferring to the Flying Corps. Medal rolls for these campaigns are in WO 100 for the Army and ADM 171 for the Navy and are held on microfilm at TNA.

Royal Engineer Balloon Section

Though the RE Balloon Section took part in the expedition in Bechuanaland in 1884/5, there does not seem to have been a medal awarded. The first medals were awarded for Balloon Section work in the eastern Sudan in 1885 and the medal roll WO 100/64 lists the name, rank and number of the men entitled to the medal.

Three balloon sections participated in the Boer War and the Queen's South Africa Medal (QSA) was awarded to men who served in the campaign up until the Queen's death on 22 January 1901. The roll for their awards is in WO 100/160 and, as well as confirming they took part in the war, the list details the clasps they were entitled to, showing which parts of the country they fought in and which battles.

Number A.G.2
Received in A.G.2
Note for Army List
Roll checked by 11/4/03
To P.O.O.
Issued

62

1st Balloon Section Royal Engineers Regiment or Corps.

ROLL of Individuals entitled to the South Africa Medal and Clasps, under the Army Order granting the Medal, issued on 1st April, 1901.

To be left blank	Regimental Number	Rank	NAME	Belmont	Modder River	Paardeberg	Dreifontein	Wepener	Johannesburg	Diamond Hill	Belfast	Defence of Kimberley	Relief of Kimberley	Defence of Mafeking	Relief of Mafeking	Cape Colony	Orange Free State	Transvaal	Rhodesia	Talana	Elandslaagte	Tugela Heights	Defence of Ladysmith	Relief of Ladysmith	Laing's Nek	Natal	Remarks
	19530	C.S.M.	Graham J.A.													Yes	Yes	Yes									
	22644	Sergt	Edwards G.A.			Yes	Yes									Yes											
	22641	"	Wellman W.J.			Yes	Yes		Yes	Yes						Yes											
	18443	"	Burton A.C.J.			Yes										Yes											
	26641	Corpl	Mellor H.													Yes	Yes	Yes									
	24404	2/Corpl	Moore W.E.						Yes	Yes						Yes	Yes										
	24488	"	Moseley A.J.			Yes	Yes		Yes	Yes						Yes											
	28130	"	Thomas S.J.			Yes	Yes									Yes											
	24464	"	Bourne W.A.			Yes	Yes		Yes	Yes						Yes											
	20888	"	Sykes H			Yes	Yes		Yes	Yes						Yes											
	1134	L/Cpl	Cormack W.				Yes		Yes	Yes						Yes											
	1218	Sapper	Barron W			Yes	Yes		Yes	Yes						Yes											
	23184	"	Cranfield N.			Yes	Yes									Yes											
	23304	"	Cooper A.													Yes	Yes	Yes									
	26923	"	Dilling A.R.			Yes										Yes											
	24499	"	Edwards C.E.			Yes	Yes									Yes		Yes									
	23247	"	Grimsditch W.H.													Yes											
	28555	"	Hillyer J.A.													Yes	Yes	Yes									
	24304	"	Hockin H.J.													Yes											
	28653	"	Ledger A.E.			Yes	Yes		Yes	Yes						Yes											

I certify that the Individuals named in this Roll were actually present at the operations for which the Medal and Clasps are claimed as above detailed,

Place Middelburg, Transvaal

Capt.

Signature and Rank of Officer personally cognizant of the Claimants' service

Part of the Royal Engineers Balloon Section Medal Roll for the Boer War. (TNA WO 100/160)

A few RE Balloon Section officers and men fought in the Boxer Rebellion (Fourth China War) and their medal roll is in WO 100/95.

The *London Gazette*

London Gazette is the government's own newspaper, published daily since 1666, which lists, among many other things, honours and awards to civilians and the military. As such it is an invaluable resource for military and family historians. Their on-line archive, which covers the period 1900 to 1997, is available at: http://www.gazettes-online.co.uk/generalArchive.asp

The archive is searchable by name (or even individual words!) so in theory it is possible to search for your relation's name and find out what medal or honour they were awarded, and when. If you have the service number try searching on that first as it will produce far fewer 'hits' than a name, even an uncommon one. Do remember though that, particularly in the First World War, service numbers were not unique. You will be presented with a downloadable PDF file of the edition(s) of the *Gazette* that contain the reference you are after. As ever, the more information you have, the more chance of finding more and because of the way the search is conducted you need to be pretty exact in what you're looking for, or to be prepared to experiment with combinations of full names and initials.

One RFC (and later RAF) officer who became famous after the war as a spy was Sidney Reilly (the 'Ace of Spies'). To find his *Gazette* references it is important to know that he was Sidney George Reilly and, using both his full name and initials, it is possible to find him listed twice in the *Gazette* of 25 January 1918 as both:

The under mentioned to be temp. Lts. (on prob.): —
2nd Sidney George Reilly. 19th Oct. 1917.

And as:

Temp. 2nd Lts. (on prob.), Gen. List, and
to be confirmed in their rank.
L. Y. Wardall. 26th Mar. 1917.
3rd Dec. 1917.
M. E. Clubine.
H. A. Kelly.
JVC. Adams.
G. W. H. Hogan.
J. H. McGee.
S. G. Reilly. [my highlight]
'Gerald Clark. , ,
•Clement Edwards.

It is possible to cut and paste information from the file onto your computer clipboard but because of the way information is copied it can be distorted in the process (though you can correct it after copying) so that the three names at the bottom of the list originally copied across as:

.S. G. Eeilly.
'Gerald dark. , ,
•Cleiment Edwards

Reilly's later award of the Military Cross for Secret Service work in Russia is in the 11 February 1919 edition of the *Gazette* and this list, interestingly, brings up two other RAF officers and an Army officer who also worked in intelligence at the same time:

Awarded the Military Cross.
Lieut. Norman Dewhurst, R. Muns. Fus.
Lieut. George Alexander Hill, 4th Bn., Manch.
R., attd. R.A.F.
2nd Lieut. .Sidney George Reilly, R.A.F.
Lieut. George Gibson Tomling, R.A.F.

The more common the name, the more references a general search will bring up so try and cut down the number of references by being as specific as you can with the range of dates you search.

As an example of some of the individuals who are listed (this time for being given temporary commissions), showing how good men could rise from the ranks, here is a further excerpt from the 25 January 1918 *Gazette*:

> 22nd Dec. 1917.
> 2nd Cl. Air Mech. Philip Henry Bayer from R.F.C.
> Actg. Flight Sjt. Jerrold Milsted, from R.F.C.
> Sjt. Mark Bernard Barrand, from R.F.C.
> 28th Dec. 1917
> Robert Leslie Finnis.
> Harry Jenks.
> Actg. Mech. William Dentith, from R.N.A.S. 30th Dec. 1917.

A useful set of tips for finding individuals (and perhaps explaining why you can't find someone you know should be gazetted) is at: http://www.military-researcher.com/LondonGazette.html. Although I have used First World War examples, the *Gazette* is just as useful for the Second World War and beyond.

Hard copies of the *Gazette* are held at TNA in their ZJ1 series, and there is a paper index in the Microfilm Reading Room there.

First World War campaign medals

The creation of the RAF while the war was still in progress created several administrative problems, and which department was liable for the issue of campaign medals, when a man had served in more than one service, was one of them. The general criteria for receipt of a medal was service in a theatre of war (the definition of which differed slightly in each service). The War Office seems to have issued the medals to men (and a few women) who had joined the RFC or RAF from other parts of the Army and who had served abroad before 1 April 1918. The Admiralty issued medals to the men who had served similarly in the RNAS.

War Office medal cards

Because the RFC was part of the Army establishment, the campaign medals that they were awarded are listed in the War Office registers, and individual cards for each officer and man are held in the WO 372 class. There are some 26,290 cards for the RFC and nearly 27,000 for the Royal Air Force and 528 Royal Naval Air Service awards are also included.

You should bear in mind, if you have original medals, that it was the practice to put the recipient's name, rank and service number around the rim, as well as the name of the unit they first served with. A man could move between units for a variety of reasons, or even between the Army, Navy and Air Force.

The cards are now available on-line from TNA website which has a search facility to allow you to search by surname and initial, as well as by service number and service that your ancestor served with. The cards were created by the Army Medal Office (AMO) towards the end of the First World War and designed to enable the AMO to place on one card all of the details about an individual's medal entitlement, their rank(s), the unit or units they served in, the first operational theatre they served in and, most importantly, the original AMO medal roll references.

All the medal cards have a covering date of 1914–20. Because of this there is no need to restrict your search by date. Please be aware that in order to narrow your search down to an individual in either the RFC, RNAS or RAF it is necessary to spell out their particular service in full, i.e. Royal Flying Corps not just RFC. If you have their service number it helps in finding the card if you put this in the 'other keywords' box. RFC and RAF service records were unique (i.e. each number applied only to one man) but the same number might be used by more than one army regiment so you may turn up more than one record and will have to decide which is your man.

Once you have identified the relevant individual you can download an image of their record card and begin to interpret the various codes upon it.

You may find that there is more than one card for an individual. This is not a mistake – cards were created for awards of gallantry medals as well as campaign medals so you may find one of these. They are useful because they give the date that the medal was gazetted.

In the case of those RFC men who were later 'remustered' to serve with the infantry (following the combing out of the Kite Balloon units to replace fitter men with lower medical categories), their War and Victory medals are marked 'RFC', as that was the unit with which they were serving when they went overseas. Any RFC men remustered to another part of the Army while in the UK, and then seeing service overseas, would receive medals marked to the regiment/corps with which they went overseas.

You can use the medal card number to find the Medal Roll Book in WO 329. These contain lists of RFC other ranks entitlements to 1914 Star, 1914–15 Star, Victory and British War Medals. RFC records are between WO 329/2300 and WO 329/2302 and in WO 329/2135, WO 329/2504, WO 329/2512, WO 329/2926, WO 329/2927, WO 329/2955, WO 329/2955, WO 329/3244 and WO 329/3272.

First World War gallantry and meritorious service awards

If you have your relative's service record you should find a mention of their campaign and any gallantry or meritorious service medals in a special section of the record usually marked 'Casualties, Wounds, Campaigns, Medals, Clasps, Decorations, Mentions' (this is how it appears on the RAF service record). There will generally be a date alongside the note, which will give you a starting point. This date is usually the date of gazette so the event for which the medal was awarded will be before that. You will need to know the unit they were serving with at the time so you can search for the recommendation that led to the award.

Recommendations for awards are usually held at unit level in AIR 1. AIR 1/163/15/142/6, for example, contains details of 3 Squadron RFC confidential reports, recommendations for awards, etc. covering the period 26 June 1917–January 1919. AIR 1/993/204/5/1216 contains Recommendations for Honours and Awards for the RFC in the Field October 1915 to August 1916.

Some recommendations are issued at Wing or Group level so it is worth looking there too. Here are two recommendations for 5 Group RAF from mid-1918:

2nd Lieutenant John McKimmie YOUNG D S M

In recognition of his skill and determination as Observer and Bomb-dropper on H.P. Machines.

Since the award of his D S M, 11/12/17 this officer has taken part as an observer and 2nd pilot in 39 raids with excellent results.

On the night of 30th May 1918, with Lieut. Russell as pilot, during the attack on La Brugeeise Works although under heavy AA fire he scored direct hits on the buildings resulting in a terrific explosion and fire. All the damage noted and reported was fully confirmed by photographs taken the following day. Two nights later he dropped bombs on Bruges Docks, with the same pilot, under heavy AA fire again scoring a direct hit on the large shed setting it on fire.

204465, HORTON, John 1st Air Mechanic (Rigger)

This man was blown into the canal during the recent bombardment, together with C W Jones 3/AME (since died), by the concussion of a large bomb bursting near.

Although severely wounded in the head this man held his comrade, who was mortally wounded, above water until help was rendered, 10 minutes later. He then collapsed through loss of blood.

A search on AIR 1 using keywords 'decorations', 'medals', 'awards' and 'honours' brings up, amongst many other records:

AIR 1/163/15/124/10: Particulars of British decorations and medals awarded to RAF personnel between 1 April 1918 and 31 December 1919

AIR 1/1767/204/142/18: Returns and correspondence on award of Service Medals and Chevrons. 1918 Jan.–Aug.

1/75/15/9/172: Honours and awards gained by officers and men. 1916–1919 (5 Group)

AIR 1/75/15/9/171: Correspondence re recommendations for honours and awards with list of those gained by officers and men. 1918 Nov.–1919 May. (5 Group)

AIR 1/113/15/39/36: Honours – award and disposal of. 1917 Feb.10–May 22. (3 Wing RNAS)

AIR 1/184/15/223/1: Roll of RAF Canadian officer recipients of honours and awards. 1919

AIR 1/727/151/1/2: Honours and award recommendation: Officers and Warrant Officers of 73 Squadron RAF 1918 July–Nov.

AIR 1/653/17/122/488: Awards for members of the RFC 1912 Aug.–1913 Mar.

AIR 1/788/204/4/613: French decorations awarded to RFC officers and men. 1914 Sept.–Dec.

AIR 1/993/204/5/1216: Recommendations for Honours and Awards. RFC in the Field. 1915 Oct.–1916 Aug.

AIR 1/1389/204/25/43: Reports on officers and recommendations for awards. 1916 Oct.–1918 Dec. (27 Squadron)

AIR 1/1898/204/225/13: Recommendations for honours and awards and nominal rolls. 1918 Apr.–1919 Jan. (148 Squadron)

AIR 1/1925/204/239/15: Honours and awards. 1918 Oct.–1919 Feb (8 Balloon Wing)

AIR 1/2147/209/3/131: Recommendations for honours and awards: HQ RFC France 1915 Oct.

AIR 1/75/15/9/173 RNAS Dunkerque Command – Honours and awards gained by officers and men. 1916–1918

Once you know your ancestor's squadron or other unit you can try searching on a combination of the unit and one of the keywords and locate the likely set of documents that will contain details of awards and commendations.

The RAF Index of Honours and Awards is still to be transferred to TNA from RAF Innsworth, when it will be held under a new reference class, AIR 81. Until it is, finding specific references to individuals can be time consuming.

RFC officers and other ranks without a service record

A good starting point is the on-line medal cards which should mention, along with the campaign medals issued, whether any other medals were issued, though apart from confirming that, say, your aircraftsman was awarded a Military Medal they are unlikely to tell you anything else. Officers are listed in the Army, Navy or RAF Lists and these should at least tell you if he was awarded, for example, a DSO and, by tracking back through the lists you can obtain an approximate date. In the event that an officer was awarded the Military Cross there are index books in WO 389 which will give you a *Gazette* date and for the Distinguished Service Order in WO 390.

RNAS officer award cards

The Microfilm Room at TNA contains original index cards for all naval officers which cover (separately) Honours and Awards; Mentioned in Despatches and Foreign Awards. Using the information provided on each card it is possible to locate a copy of the original citation

Foreign awards

Foreign awards to British officers and men were frequently referred to the Foreign Office so that a search in the Foreign Office card index can find references that will you lead you to the papers in the FO 372 (Treaty) series. Though there is usually little detail given they are worth looking at for the sake of completeness.

Many Canadians served in the RFC, RNAS and RAF during the war and were awarded British medals. The Royal Canadian Air Force has a website dedicated to these men at http://www.airforce.ca/index2.php3?page=wwi which is alphabetical and contains much personal detail, as well as details of the medals awarded.

Medals after the First World War

Medal records from the early 1920s up until the mid-1950s (though a very few, probably released in error, go up to the early 1970s, including records for RAF crews involved in the early troubles in Northern Ireland) are in AIR 2 under Code B, 30, DECORATIONS, MEDALS, HONOURS AND AWARDS.

There are some 995 files in this series relating not only to the issue of medals to individuals but to questions such as to what they could be awarded for, how they are to be worn and whether foreigners could be awarded them. Only occasionally are individuals named in the index – AIR 2/119: 'Decorations to Capt. Ross Smith and Crew for accomplishing First Flight to Australia' and AIR 2/318: 'Award of A.F.C. to Lt.R.R. Bentley, M.C. South African Air Force in recognition of his flight from England to Cape Town. (1927)' are two of these.

Campaign medals between the wars

The RAF took part in several small campaigns in the Middle East, India and Africa between the wars, and the usual campaign medal would be a 'General Service' medal with a bar for the specific campaign. Thus for the Somaliland campaign of 1920 the RAF contingent were awarded the Africa General Service Medal with a bar marked 'Somaliland 1920'. For operations in India men were awarded the Indian General Service Medal and in Iraq either the general service medal with a bar or the Iraq Active Service Medal.

AIR 2 records relating to campaign medals are:

AIR 2/204: Recommendations for honours and awards for 'Z' Expedition in Somaliland

AIR 2/2267 to AIR 2/2270: Africa General Service Medal with clasp 'Somaliland 1920' – claims, grants and awards

AIR 2/473: Grant of Iraq Active Service Medal in connection with operations in Iraq

AIR 2/632: Grant of Indian General Service Medal in connection with operations on the North West Frontier, 1930–1931.

AIR 2/448: Recommendations for honours and rewards in connection with operations on the North-West Frontier (India), 1930–1935

AIR 2/9393: Immediate awards: Waziristan operations 1938

Gallantry and other awards between the wars

Recommendations and awards for bravery, distinguished flying and for distinguished service are also in AIR 2 series and include:

AIR 2/1581 and AIR 2/1582: Order of the Bath: submissions 1919–1945

AIR 2/1727: Miss Jean Batten: suggested award of an honour in recognition of record flight

AIR 2/2084: Recommendations for the award of the Victorian Order: RAF Display 1937

AIR 2/2489: New Year Honours List: awards for flying services in peacetime 1937 to 1960

AIR 2/9315: New Years Honours List, 1939: rewards for valuable flying services in peacetime

AIR 2/10198: Appointments to Military Division British Empire Order, Kings Birthday, June 1926

AIR 2/10207: Rewards for valuable flying services under peacetime conditions: New Year Honours List 1931

Second World War campaign medals

There are ten campaign medals for the Second World War, though the rules mean that no individual is entitled to wear more than five. An individual with the maximum five medals was allowed a 'clasp' to wear on one of his medal ribbons to show service, which would have entitled him to another medal.

A (much simplified) list of medal entitlements is as follows.

The 1939–45 War Medal

Awarded for all servicemen who served for 28 days during the war. Awarded to all three services.

The 1939–45 Star

Awarded to servicemen who saw active service overseas. Awarded to all three services. There is a special clasp 'Battle of Britain' for members of the 61 RAF fighter squadrons who took part in the battle. This was the only medal awarded for service in France 1939–40, Norway, Greece, Crete and on commando raids.

Africa Star

For service in East Africa or North Africa and Malta 1940–3. There are bars for service with 8th Army, 1st Army and North Africa 1942–3. Awarded to all three services.

Air Crew Europe Star

For all aircrew who flew over occupied Europe between 1939 and 5 June 1944. If they were later entitled to the France and Germany Star they wore a clasp on this medal instead. RAF only.

Atlantic Star

Awarded mainly to Royal and Merchant Navy personnel for service in the Atlantic 1939–45, though some RAF and Army personnel were entitled.

Burma Star

For service in the Burma Campaign 1941–5 but not for service in Malaya in 1941 for which the Pacific Star was awarded. Awarded to all three services.

France and Germany Star

For service in North West Europe from 4 June 1944 to 8 May 1945, including Belgium and Holland. Awarded to all three services.

Italy Star

For service in Italy and Sicily 11 June 1943 to 8 May 1945 but also awarded for service between these dates in Greece, the Aegean, the Dodecanese, Corsica, Sardinia, Yugoslavia and southern France. Awarded to all three services.

Pacific Star

Awarded for service in the Far East (excluding Burma) including Malaya and Singapore (1941–2), Hong Kong (1941) and the reconquest of Japanese-occupied territory. Awarded to all three services.

Defence Medal

Awarded for the Defence of Great Britain to all three services, the Home Guard, Civil Defence and Medical Services. Also awarded for service in Ceylon, West Africa, Malta, Cyprus and Gibraltar.

Unlike medals for First World War, details of recipients were not engraved on the medals. Please note that the serviceman (or woman) did not receive their medal automatically, unless they were still serving at the time of issue, but very many of them would have already been discharged by this stage. They were expected to apply for them, and notices were posted in the press. Many thousands did not make the claim, so if you can't find medals don't assume that your relative didn't serve.

Medals can still be claimed by veterans who didn't claim them at the time, and replacements can be claimed by contacting: Armed Forces Personnel Administration Agency, Building 250, RAF Innsworth, Gloucester GL3 1HW. Email: JPAC@afpaa.mod.uk; fax: 0141 224 3586; free phone: 0800 085 3600. Details of how to apply can be found at the Veterans Agency website at: http://www.veteransagency.org.uk.

Records relating to the issue of Second World War campaign medals are still retained by the Ministry of Defence.

Second World War gallantry and distinguished service awards

Recommendations for Second World War awards are, once again, in the AIR 2 series. Not only are there recommendations for serving personnel but also for civilians involved in the air services or events involving aircraft (such as accidents and crashes). There are awards for instructors, bomb disposal, test flying and to prisoners of war.

To flying personnel awards fell into two categories, immediate awards (for individual acts of bravery) and non-immediate (for extended periods such as bomber tours of operation). The recommendations are grouped by command (Bomber,

Fighter, Coastal, Middle East, etc.) and chronologically. Some typical examples for Bomber Command include:

AIR 2/4072: Recommendations for awards to RAF personnel: Bomber Command operations 1939–1940

AIR 2/4094: Immediate awards to RAF personnel: Bomber Command operations 1939–1940

AIR 2/6085: Bomber Command non-immediate awards 1940–1941

AIR 2/8748: Non-immediate awards: Bomber Command 1945–1946

AIR 2/9447: Immediate awards: Bomber Command 1940 July–Aug.

Fighter Command awards covering the Battle of Britain are in:

AIR 2/4086: Recommendations for honours and awards (non-immediate): Fighter Command operations

AIR 2/4095: Immediate awards to RAF personnel: Fighter Command operations

AIR 2/8351: Non-immediate awards: Fighter Command, Sept. 1940

AIR 2/9468: Non-immediate awards: Fighter Command 1940 Aug.–Sept.

Immediate awards tend to be recommended across the RAF as a whole so if you don't know why an award was made (and many, including the Distinguished Service Order and the Distinguished Flying Cross could be awarded for both individual acts of gallantry and for long periods of bravery) you may need to check several files, even if you know the date of gazette. There are thirteen files of recommendations for 1943 alone!

Awards for ground gallantry include awards to soldiers who assisted the RAF, as well as to RAF Regiment soldiers and to airmen serving at aerodromes.

Examples of the kind of recommendation included are AIR 2/8908:

NOTABLE WAR SERVICE – 1358834 AC1 JARVIS RAF STATION LEIGHTON BUZZARD

1. The Commander in Chief desires to bring to the attention of all ranks the courage displayed by the above named airman in the following circumstances.
2. At approximately 2100 hours on the night of 22nd November 1941 an aircraft crashed at Brabazon Road, Norwich, setting fire to several houses. AC1 JARVIS who was on leave in Norwich at the time, immediately rushed to the scene of the crash and joined civil defence workers and others (including the crew of the aircraft) who were engaged in rescue work.
3. Showing great personal courage and a complete disregard for his own safety, AC1 JARVIS entered the burning houses and successfully brought out an injured woman who would otherwise have burned to death.

AIR 2/4900, 'DECORATIONS, MEDALS, HONOURS AND AWARDS (Code B, 30): Immediate awards: operational Commands 1942' contains a recommendation for:

P/O Richard Millen Horsley – First Wireless Operator He was posted to the Squadron w.e.f. 16/8/41 and he has carried out 27 trips without a single failure. He has always been of the greatest possible assistance to his crew. His trips have included attacks on many important targets in Germany and have included attacks on HAMBURG and many flights to the RUHR. He also took part in the combined operation against VAAGSO ISLAND when he machine gunned the ground defences effectively while his aircraft was engaged in laying a smoke screen on the island. He always set an example of cheerful courage in the face of the enemy. This officer is one of five who escaped to England.

P/O Horsley's conduct during the action at Cologne was a fine example to the remainder of the crew. He showed great coolness and skill in removing from the turret, Sgt Naylor, who had been wounded and in treating his wounds during a time of great stress and danger. I strongly recommend that he be granted an immediate award of the Distinguished Flying Cross.

AIR 2 also includes recommendations to civilians and other serviceman recommended by the Air Ministry for awards for assisting the RAF including, on one occasion, a schoolboy recommended for kicking through a glass door to rescue a baby trapped by a crashed aircraft. He received a commendation, as well as two days off school with lacerated feet!

A professional researcher, Paul Baillie, has indexed all the recommendations within AIR 2. He can be contacted via email at paulbaillie@tiscali.co.uk. For a very reasonable fee he can check his index and provide you with a copy of any surviving recommendation.

Recommendations to the Sovereign

The AIR 30 files contain submissions to the Sovereign for royal approval of appointments, promotions, awards and regulations, and petitions in court martial cases. They are chronological, so if you know the date of an award you can look for the recommendation that prompted it. The following are both from AIR 30/46

Awarded a Bar to the Distinguished Flying Cross
Flying Officer DUDLEY LLOYD EVANS MC DFC RAF
(Original Distinguished Flying Cross Gazetted 3rd December , 1918)

For gallantry, skill and devotion to duty on the 1st November, 1920, whilst accompanying another machine on reconnaissance. Owing to engine trouble, the other machine, with pilot and observer, had to make a forced landing in hostile country. A party of mounted Arabs at once started firing at the observer who was dismantling a Lewis gun. On seeing this, Flg Officer Evans landed at great peril to himself, took both officers on his already loaded machine and, getting off with much difficulty, returned to Headquarters.

313619 Flt Sgt Frank Bebbington

For consistent good work and devotion to duty. By his example, tact and organisational powers, he is largely responsible for the output of machines at Aircraft Park during operations. This NCO has frequently volunteered to work all night in addition to his day work, when circumstances have arisen.

Air Ministry Bulletins from the Second World War are held at the RAF Museum and give details of medal citations for RAF servicemen. These often include a small amount of information on a man's (or woman's) previous career for the press who might be interested in background of a local hero. Amongst the citations is one for Sgt Donald Benjamin Godfrey who was awarded the George Medal. He was the rear-gunner of a Wellington Bomber, which crashed in bad visibility in December 1942. He extricated himself from the burning wreckage, hauled the injured wireless operator to safety, helped the pupil pilot get clear and then realized that the instructor was still trapped. He dashed back to help.

> A petrol tank exploded and heavy pieces of wreckage were hurled into the air, but this did not deter Sergeant Godfrey and, in spite of the heat and flames he hacked and pulled at the wreckage until finally he succeeded in releasing the instructor. He was pulling him clear when assistance arrived. Sergeant Godfrey sustained severe burns causing temporary blindness. He displayed courage and devotion to duty in keeping with the finest traditions of the Royal Air Force.

Post-Second World War awards

Once again, AIR 2 holds those records which have been released relating to RAF awards, though the files are sparse and there appear to be a large number of gaps. Files that may be of interest include:

AIR 2/12276: Awards by the USA to RAF and Dominion Air Forces in Korea

AIR 2/16815: Awards for services in Korea: recommendations; December 1950 onwards

AIR 2/12423: Awards for service in Kenya 1953–1955

AIR 2/12424: Awards for service in Kenya 1955–1956

AIR 2/12430: Africa General Service Medal award for operations in Kenya

AIR 2/13941: Award of cards for good service: introduction of system for operations in Malaya

AIR 2/14043: Award of General Service Medal for service in Malaya since 16 June 1948

AIR 2/18214: Queen's Birthday Honours List 1968: flying awards

AIR 2/18268: New Years Honours List 1969: flying awards

AIR 2/18257: RAF awards for gallantry on the ground 1968–1971

AIR 2/18258: RAF awards for gallantry in the air 1967–1970

AIR 2/19194: Awards in operational lists: recommendations, Northern Ireland November 1971–April 1973

AIR 2 Files continue to be released under the thirty-year rule.

Casualties and accidents

If you believe, or know, that your relative died or was killed after the beginning of First World War, the best place to start looking for information is the Commonwealth War Graves Commission (CWCG) website at http://www.cwgc.org. This has a searchable database of casualties and we noted Lieutenant Guy's details when looking at the RFC in First World War. Reggie Heath who, as we saw, died as a prisoner of the Japanese in 1943 is also recorded, this time as being commemorated on the Singapore Memorial because his body was never recovered.

It is possible to search using just surname and initials but the more detail you are able to provide the closer the selection of results you get will be to the person you are looking for. Please note that, even though the RFC came under the Army, and the RNAS under the Royal Navy, you'll need to search under Air Force in order to get a result!

The CWGC website only contains records up to, and including, 1947.

First World War casualties

If you know from the CWCG website, or from the service record, the unit to which your ancestor was attached at time of death it is worth searching AIR 1. We have already seen with Lieutenant Guy that several documents had to be located, including the ORB and combat reports, in order to obtain a picture of his death. For many other units there are lists of casualties, varying in detail from simple lists to detailed explanations. Try searching on a combination of squadron number and 'casualty' or simply on the squadron number and look at the records that are brought up. A search on 'casualty' will also bring up files indexed under 'casualties'. AIR 1/2011/204/305/11 contains the casualty returns for 11 Squadron between September 1917 and March 1919 for example.

RFC/RAF casualty cards

The RAF Museum at Hendon holds RFC and RAF casualty cards. The earliest cards are for the RFC only but after 1 April 1918 they include the officers who transferred from the RNAS, and run up until about 1928. Created when a man was injured, sick or killed, they give details of what was wrong with him and some information about the circumstances in which it occurred. The RFC's earliest casualty was Sergeant Major D S Jillings who is normally reported as having been shot in the 'upper thigh' on 22 August 1914. His casualty card records 'G.S.W. (gunshot wound) in buttock'. He was back with his squadron by 8 September but his wound obviously continued to bother him as he took sick leave in 1915 with neuralgia of the spine resulting from his previous wound. The casualty cards can be particularly useful if a relative was taken as a prisoner of war, as records for these are generally sparse. Lieutenant Colin Elphinstone of 21 Squadron RFC was shot down on 15 September 1916 during the Battle of the Somme. The card was originally created to record him as missing but his mother received a letter in November stating that he was wounded and a prisoner. Subsequently the Foreign Office also reported him as a POW and that he was suffering from GSW right foot and wasting away of left lower arm. The American Embassy, which looked after British subjects in Germany, also confirmed his capture. It was only in May 1918 that the Red Cross confirmed he was being held at Holzminden Camp. At the war's end he was in Halt Camp in Jutland and was repatriated in December 1918, landing at Leith.

RNAS casualties

As part of the Royal Navy, casualties for the RNAS are listed in ADM 242/7 to ADM 242/10. They are compiled alphabetically and include all ranks. They give ranks, dates of birth and death, place of death and details of next of kin. There are index cards for RN officers only between ADM 242/1 to ADM 242/6 which give name, date and place and cause of death.

Casualties between the wars

The RAF casualty cards (see above) continue to 1928. Unit records such as ORBs should generally mention casualties, particularly officers. Casualties in combat were rare and are also likely to be mentioned in the *London Gazette* if a supplement was produced for the campaign. AIR 2/11613 contains details of a 'Court of Enquiry and associated papers into a forced landing of aircraft DH9a H147 at Jabilah, South Iraq, on 24th July 1924, involving F. Lt. WC Day and P O DR Stewart leading to their eventual death'.

Courts martial

Almost all servicemen go through their career without committing any serious offences, the vast majority of disciplinary cases being dealt with by the serviceman's commanding officer and resulting in a stoppage of pay for a couple of days. A small number, inevitably, were charged with offences resulting in trial. Curiously enough, if this is the case, though their service records may be retained, some of their court martial records may actually be available, right up to the 1960s!

Courts martial are the military (including the RAF) courts for trying servicemen (and some civilians abroad) for crimes committed whilst under military jurisdiction.

The service record will give you certain basic information if your relative was tried in a military court. The section 'Time Forefeited', i.e. time in the service that does not count towards a pension, will tell you if they lost time because of a court martial sentence and will give you the date of the trial. This is all you need to get started. Though most court martial papers are lost or still retained there are registers available which will give you basic details. The RAF inherited its court martial system from the RFC and the systems they used were similar.

There are two main types of court martial: the District Court Martial, which can only try cases with a sentence of less than two years, and General Courts Martial, which try the most serious crimes. Officers are always tried by General Courts Martial. A third category, Field General Court Martial, applies during wartime and can impose the death penalty provided the three officers required to be present were unanimous.

RFC courts martial records, because they came under the army, are in the War Office records. Records of Army Field General Courts Martial covering First World War are between WO 71/387 and WO 71/679. These have been indexed on-line and so can be searched for on TNA search engine.

Registers of charges between 1880 and April 1918 (thus covering the Royal Engineers Balloon Sections as well as the RFC) are between WO 86/29 and WO 86/81.

WO 213/1 to WO 213/21 contain registers of Field General Courts Martial and military courts between 1909 and April 1918.

RNAS courts martial records are in the ADM papers. ADM 194/43 to ADM 194/45 contain registers of courts martial 1857 to 1915 and will provide basic details of trials and verdicts within that period. ADM 156 contains courts martial cases and files 1890 to 1965, ADM 178 contains naval courts martial cases, boards of inquiry reports, and other papers. Very few records of courts martial are in these files, but you should be able to get a reasonable picture of the charge and verdict from the information on the service record.

An interesting case that does survive concerns allegations that RNAS officers were involved in looting in the shellfire-damaged Belgian town of Ypres in 1915. ADM 156/79 contains the complete transcript of the court martial of Lieutenant Commander Percival Parry RNVR and certain other naval officers attached to the RNAS. It was alleged that, against orders, Parry had taken his armoured car into the city and looted some vestments from the damaged cathedral. Having considered the evidence of the officers involved, and various RNAS ratings who drove Parry's armoured car, the court decided that he was not guilty. In addition, they came to the conclusion that the whole accusation had been made up as part of an elaborate plot to discredit Parry and have him replaced by another officer. Two senior officers were admonished for their part in the affair. Squadron Commander Longmore

> is quite unfitted for the responsible position he now holds, and on a full review of his conduct . . . it would be to the advantage of His Majesty's Service if Wing Commander Longmore could be relieved of his present position and sent back to sea service.

Lieutenant Commander Chilcott's services were 'dispensed with' in order to avoid unnecessary adverse publicity to the Air Service.

One thing to bear in mind, especially if your ancestor served on naval airships, is that the loss of a ship automatically results in a court martial for the officer in command. Aeroplanes didn't count as ships, so could be crashed with (legal) impunity. Airships were treated as ships so that the loss or damage to one (including the all too frequent ripping of the balloon envelope) would result in a hearing.

There are a variety of records available relating to specific RAF courts martial. AIR 21 contains registers of charges giving the name and rank of each prisoner, place of trial, nature of the charge and sentence. Held in the form of ledgers, these record the date and place of the court, the name, rank and unit of the accused, as well as basic details of the offence and sentence. Charges listed are: desertion, absence, striking or violence, insubordination or disobedience, leaving post, drunkenness, theft, fraud, cheque fraud, indecency, resisting escort, escaping confinement, scandalous conduct, self-inflicted wound, miscellaneous. Miscellaneous charges include specific offences against King's Regulations – the number quoted in the column refers to the specific regulation breached. AIR 10/2279 is a copy of The Kings Regulations and Air Council Instructions for the Royal Air Force that should tell you the nature of the offence.

The nature of the charge is indicated by 'Do' in the relevant column. In addition to the basic charges you may find additional letters in the column that will help flesh out the nature of the alleged offence. These are:

ABH	Actual Bodily Harm
CA	Common Assault
WI	Wounding with intent
W	Wounding
SB&L	Shopbreaking and larceny
HBL	Housebreaking and larceny
B & E	Breaking and entering
WD	Wilful damage
MFD	Making false document
POSB	Post Office Savings Book
Forgery	Forgery
R (in theft column)	Receiving
FP	False pretences

For offences of indecency 'M' or 'F' indicates the sex of the person with/on whom the offence was committed. Miscellaneous military offences were noted under the section number of the offence under King's Regulations, so that 69 was a military offence and 56 was for escaping. If the accused was acquitted of a charge a red ring was put around the 'Do' in the charge column.

As the AIR 21 record will tell you the date and place of the court martial, it is always worth checking the local press for reporting of the case or, if sufficiently serious, the national press. Journalists have always loved a trial and scandal!

AIR 18 files contain records of the Judge Advocate General's Office relating to the courts martial of RAF personnel from 1941 to 1994. Inevitably later records remain closed under the thirty-year rule, and some documents are retained even in the open records because they contain particularly sensitive information that the law does not allow to be released.

Typical documents that you will find in the AIR 18 files include: summaries of evidence, witness statements to the court, lists of exhibits, descriptions of the exhibits, service record summaries, lists of witnesses, statements given to the police, transcripts of the trial, records of proceedings naming the members of the court martial and defending and prosecuting counsels, legal notes on the proceedings by the Advocate General.

Please bear in mind that many participants in courts martial in AIR 18 are likely to still be alive and take care in using the information you may find in the records.

AIR 43 records comprise the Judge Advocate General's Office: Royal Air Force Courts Martial Charge Books covering 1918 to 1948. These consist of the opinions of the Judge Advocate General's Department on evidence to be used in various cases, on the nature of the charges to be pressed and the procedure. Only occasionally is a verdict recorded. The files are chronological so if you know from their service record that a court martial was held you can check the relevant AIR 43 file to see if any papers are there. The files are not indexed but do appear to be scrupulously in date order so it shouldn't take long. For many cases, apart from the record in AIR 21, the AIR 43 papers will be the only official papers remaining.

AIR 71 records are administrative papers (Out letters) relating to courts martial and are bound chronologically so if you know the date of the trial you can locate any relevant documents.

An RAF court martial

In 1923 Reginald Bone was a Wing Commander and commanding officer of RAF Calshot. A junior officer, Observer Officer Rowland Nicholson, was charged with disobedience (two charges) and insubordination. Though the proceedings of the court martial are not available, it is possible to get some picture of the proceedings and charges.

AIR 21/3 lists, on p. 11 of the ledger, show that a court martial was held at RAF Calshot on 17 July 1923 and that charges of Disobedience and Insubordination were brought against him by the Wing Commander himself. He was found not guilty of the second charge of disobedience (a red ring appears around the charge) but found guilty of insubordination and disobedience on the first charges and sentenced to be dismissed the service.

AIR 71/2 unfortunately adds little information, merely confirming that the court martial took place on 17 July 1923 at Calshot, gives some details as to the arrangements for Mr Sutherland Graeme to act as Judge Advocate, and that copies of the proceedings were sent to the Judge Advocate General's office on 21 July.

AIR 43/3 gives a little more detail of the incidents that led to the trial. The charge of using insubordinate language to his superior officer resulted from an incident on 11 June 1923.

> During an official investigation by Wing Commander Bone, C.B.E., D.S.O., his Commanding Officer concerning certain telephone messages sent by him (the accused) said in an insolent manner in reply to a question put to him by Wing Commander Bone, 'That is correct; the signals officer of the Isle of Grain is posted to the Argus and does not want to go he has been posted to seventeen different stations in the last three years, so you may take that down if you want to make a song about it', or words to that effect.

The second charge 'Disobeying a lawful command given by his superior officer' relates to an order given to Observer Officer Nicholson that he should not leave the station that night, but that he did so.

AIR 44/3 gives a little more detail about the further disobedience charge, in that he was ordered, in writing, by Wing Commander Bone, to return to Calshot Air Station by the 18.20 boat from Southampton but failed to do so. The document confirms the verdict of guilty to the insubordination and first disobedience charges and confirms that the charges were 'well laid' that there was sufficient evidence to justify the findings and that the sentence was according to the law.

Though he was sentenced to dismissal from the service, AIR 21 also shows that his sentence was commuted to a severe reprimand, though the RAF List shows that he retired from the RAF only two months later, on 15 September 1923. Perhaps some deal was done, or perhaps the shame of the trial and being found guilty persuaded him to go.

Please be aware that courts martial are serious affairs and that many relatives may be offended to have their details known about, even though the documents are in the public domain. Even what we might now consider to be a relatively trivial offence might have carried stigma in the past and relatives might not be keen to discuss either their own personal involvement or that of their father or other close relation.

Appendix 1

RANKS IN THE RFC/RNAS/RAF

Commissioned ranks RNAS/RFC/RAF

RNAS 1912–18	*RFC & early RAF*	*RAF 1919 onwards*
		Marshal of the Royal Air Force
	General	Air Chief Marshal
	Lieutenant General	Air Marshal
	Major General	Air Vice Marshal
	Brigadier	Air Commodore
Wing Captain	Colonel	Group Captain
Wing Commander	Lieutenant Colonel	Wing Commander
Squadron Commander	Major	Squadron Leader
Flight Lieutenant	Captain	Flight Lieutenant
Flight Sub-Lieutenant	Lieutenant	Flying Officer
	Second Lieutenant	Pilot Officer
	Acting Pilot Officer	

WRAF 1918–20	*WAAF and WRAF 1939–67*
Commandant	Air Chief Commandant (equivalent to Air Vice Marshal)
Deputy Commandant	Air Commandant
Assistant Commandant I	Group Officer
Assistant Commandant II	Wing Officer
Administrator	Squadron Officer
Deputy Administrator	Flight Officer
Assistant Administrator	Section Officer
	Assistant Section Officer

Non-commissioned officers and airmen

RNAS	*RFC*	*WRAF 1918–20*
Chief Petty Officer I	Warrant Officer	Senior Leader
Chief Petty Officer II	Quartermaster Sergeant	Chief Section Leader
Chief Petty Officer III	Flight Sergeant	Section Leader
Petty Officer	Sergeant	
Leading Mechanic	Corporal	Sub Leader
Acting Air Mechanic I	Air Mechanic I	
Air Mechanic II	Air Mechanic II	
	Air Mechanic III	Member

RAF 1918	*RAF 1918–51*
Sergeant Major I/Chief Master Mechanic/Chief Master Clerk	Sergeant Major Class 1
Sergeant Major II/Master Mechanic/ Master Clerk	Sergeant Major Class 2
Flight Sergeant/Chief Mechanic/ Flight Clerk	Flight Sergeant
Sergeant/Sergeant Mechanic/Sergeant Clerk	Sergeant
Corporal/Corporal Mechanic/Corporal Clerk	Corporal
Air Mechanic 1st Class/Clerk 1st Class	Leading Aircraftsman
Private 1st Class/Air Mechanic 2nd Class/Clerk 2nd Class	Aircraftsman 1st Class
Private 2nd Class/Air Mechanic 3rd Class/Clerk 3rd Class	Aircraftsman 2nd Class

Appendix 2

ROYAL FLYING CORPS ORDER OF BATTLE

The RFC in France, it must be remembered, was part of the Army, and its organizational structure (its 'order of battle') reflected its role supporting the various armies that were created as the war continued and the size of the army increased.

Initially there was just the British Expeditionary Force (BEF), consisting of two infantry corps and the cavalry division. As the war progressed the BEF expanded to consist of five armies, each of which had an RFC Brigade attached to it.

The initial order of battle, for August 1914, was simple:

RFC	2nd Aeroplane Squadron
	3rd Aeroplane Squadron
	4th Aeroplane Squadron
	5th Aeroplane Squadron
Line of Communication Units	1 Aircraft Park

By March 1915, after the separation of the original BEF into First and Second Armies the Order of Battle became:

FIRST ARMY

RFC	*First Wing*
	Headquarters
	2nd Squadron
	3rd Squadron

SECOND ARMY

RFC	*Second Wing*
	5th Squadron
	6th Squadron

GHQ TROOPS

RFC	*HQRFC*
	1st Aeroplane Squadron 4th Aeroplane Squadron 9th Aeroplane Squadron Aircraft Park (Advanced Echelon)

By August 1915 the order of battle was even more complex:

GHQ TROOPS

RFC	*HQRFC*
	7 Squadron 1st Aircraft Park (Advanced Echelon)

Army Troops

FIRST ARMY

RFC	*First Wing*
	Headquarters 2 Squadron (attached IV Army Corps) 3 Squadron (attached I Army Corps) 10 Squadron

16 Squadron (attached III Army Corps and Indian Army Corps)
No. 6 Naval Kite Balloon Section)

SECOND ARMY

	Second Wing
	Headquarters 1 Squadron 5 Squadron 6 Squadron No. 2 Naval Kite Balloon Section (attached II Army Corps) No. 4 Naval Kite Balloon Section

THIRD ARMY

RFC	*Third Wing*
	Headquarters 4 Squadron (attached X Army Corps) 8 Squadron (attached VII Army Corps) 11 Squadron 3rd Aircraft Park (Advanced Echelon)

Corps Troops

I ARMY CORPS

RFC	3 Squadron (attached)

In August 1915 the structure changed so that each army was supported by an RFC brigade consisting of two aeroplane wings and a balloon wing, and this was the structure in place at the end of the war.

As can be seen, the various squadrons were not fixed as part of any particular wing or even army so that to find records that belong to them as part of the records of a higher echelon (such as a wing or a brigade) you'll need to check the RFC order of battle for the period you're interested in. There's a comprehensive list of orders of battle in AIR 1/2129/207/83/1.

RFC order of battle in Italy, January 1917

7th Brigade RFC
Headquarters, 14th Wing
Headquarters, 51st Wing
28, 34, 42 and 66 Squadrons
7th Aircraft Park
Headquarters, 9th Balloon Company
Headquarters, 20th Balloon Company
Headquarters, 4th Balloon Wing
Nos. 7, 22, 33 and 34 Kite Balloon Sections.

Appendix 3

RNAS ORDER OF BATTLE, 31 MARCH 1918

Overseas units

1 Wing (Dunkirk)	*2 Wing (Mudros/Aegean)*
2 Squadron 13 Squadron 17 Squadron	A Squadron (based at Thasos) B Squadron (based at Miylene) C Squadron (based at Imbros) D Squadron (based at Stavros) E Flight (based at Hazni Junas) F Squadron (based at Mudros) G Squadron (based at Mudros) Z (Greek) Squadron (based at Thasos)
4 Wing (La Panne)	*5 Wing (Malo-les-Bains)*
7 Squadron 11 Squadron 12 Squadron	4 Squadron 8 Squadron
6 Wing (Otranto)	*Overseas Kite Balloon Stations*
1 Seaplane Squadron 2 Seaplane Squadron Aeroplane Flight Wing Equipment Base	Alexandria, Bizerte, Brindisi, Corfu, Gibraltar, Malta

Units with RFC in France

13 Wing, 3 Brigade
1 Squadron
10 Wing, 1 Brigade
10 Squadron
3 Squadron
22 Wing, 5 Brigade
5 Squadron

11 Wing, 2 Brigade
6 Squadron
9 Squadron
41 Wing, 8 Brigade
16 Squadron

Overseas Seaplane Stations

Alexandria
Cherbourg
Gibraltar
Malta
Mudros
Otranto
Port Said
Suda bay
Syra Island
Taranto

Overseas Airship Station

Mudros

RNAS bases in Great Britain

Aldburg (Suffolk)
Ashington (Northumberland)
Bacton (Norfolk)
Bembridge (Isle of Wight)
Brading (Isle of Wight)
Butley (Suffolk)
Calshot (Hampshire)
Cattewater (Devon)
Chingford (Essex)
Cramlington (Northumberland)
Cranwell South (Lincolnshire)
Dover Harbour (Kent)
Dundee (Forfar)
Eastchurch (Kent)
Fairlop (Essex)
Fishguard (Pembrokeshire)
Gosport (Hampshire)
Greenland Top (Lincolnshire)
Hendon (Middlesex)Holt (Norfolk)
Anglesey/Llangefni
Atwick/Hornsea (Yorkshire)
Bangor (Caernarvonshire)
Bembridge Harbour (Isle of Wight)
Burgh Castle (Norfolk)
Cairncross (Berwickshire)
Catfirth (Shetlands)
Chickerill (Dorset)
Covehithe (Suffolk)
Cranwell North (Lincolnshire)
Donibristle (Fife)
Dover/Guston Road (Kent)
Eastbourne (Sussex)
East Fortune (East Lothian)
Felixstowe (Suffolk)
Freiston (Lincolnshire)
Grain & Port Victoria (Kent)
Hawkcraig Point (Fife)
Hornsea Mere (Yorkshire)

Houton Bay (Orkney)
Killingholme (Lincolnshire)
Lee on Solent (Hampshire)
Leuchars (Fife)
Levington (Suffolk)
Leysdown (Kent)
Luce Bay (Wigtown)
Macrihanish (Argyll)
Manston (Kent)
Martlesham Heath (Suffolk)
Mullion (Cornwall)
New Haggerston (Northumberland)
Newhaven (Sussex)
Newlyn (Cornwall)
North Coates Fitties (Lincolnshire)
Owthorne (Yorkshire)
Padstow (Cornwall)
Pembroke (Pembrokeshire)
Pierrowal (Orkney)
Portholme Meadow (Huntingdonshire)
Portland Harbour (Dorset)
Prawle Point (Devon)
Redcar (Yorkshire)
Rennington (Northumberland)
Rosyth (Fife)
Scapa (Orkney)
Seahouses (Northumberland)
Seaton Carew (Durham)
Seaton Carew/Tees (Durham)
Smoogroo (Orkney)
South Shields (Durham)
Stenness Loch (Orkney)
Stonehenge (Wiltshire)
Strathbeg Loch (Aberdeenshire)
Tallaght (Co Dublin)
Telscombe Cliffs (Sussex)
Torquay Harbour (Devon)
Tregantle (Cornwall)
Tresco (Isles of Scilly)
Turnhouse (West Lothian)
Tynemouth (Northumberland)
Walmer (Kent)
Westgate (Kent)
Westward Ho! (Devon)
Withnoe (Cornwall)
Yarmouth (Great) (Norfolk)

Appendix 4

RAF ORDER OF BATTLE IN FRANCE, 11 NOVEMBER 1918

HQRAF

Headquarters Communication Squadron
No. 1 Aircraft Depot (includes 1 Port Depot)
No. 1 Aircraft Depot (D)
No. 1 Aircraft Depot (M)
No. 1 Aeroplane Supply Depot
No. 2 Aircraft Depot (Includes 1 Port Depot)
No. 2 Aeroplane Supply Depot
Engine Repair Shops
British Aeronautical Supplies Department

9th (GHQ) Brigade

9th Wing: 18, 25, 27, 32, 49, 62 Squadrons
51st Wing: 1, 43, 94, 107, 205 Squadrons
54th Wing: 63, 151, 207 Squadrons
82nd Wing: 58, 152, 214 Squadrons
9th Aircraft Park
5th Air Ammunition Column
9th Air Ammunition Column
6th Reserve Lorry Park
20th Reserve Lorry Park

1st Brigade (attached 1st Army)

1st Wing: 5, 16, 52 Squadrons; L Flight
10th Wing: 19, 22, 40, 64, 98, 148, 203, 209 Squadrons; I Flight
1st Balloon Wing: 12, 4, 10 Companies
1st Aircraft Park
1st Air Ammunition Column
9th Aircraft Park
5th Air Ammunition Column
9th Air Ammunition Column
6th Reserve Lorry Park
20th Reserve Lorry Park

2nd Brigade (attached 2nd Army)

2nd Wing: 4, 7, 10, 53, 82 Squadrons; M Flight
11th Wing: 29, 41, 48, 70, 74, 79, 149, 206 Squadrons
65th Wing: 38, 65, 108, 204 Squadrons
2nd Balloon Wing: 5, 6, 7, 8, 17 Companies
2nd Aircraft Park
5th Aircraft Park
8th Aircraft Park
9th Aircraft Park
5th Air Ammunition Column
9th Air Ammunition Column
2nd Reserve Lorry Park
7th Reserve Lorry Park
2nd Air Ammunition Column
7th Air Ammunition Column
No. 8 Salvage Section

3rd Brigade (attached 3rd Army)

12th Wing: 12, 13, 15, 59 Squadrons; N Flight
13th Wing: 56, 60, 87, 201, 210 Squadrons
90th Wing: 38, 65, 108, 204 Squadrons
3rd Balloon Wing: 12, 16, 18, 19 Companies
3rd Aircraft Park
3rd Air Ammunition Column
3rd Reserve Lorry Park
19th Reserve Lorry Park
No. 6 Salvage Section
No. 9 Salvage Section

5th Brigade (Attached 4th Army)

15th Wing: 6, 8, 9, 35, 73 Squadrons; 3rd Squadron AFC
22nd Wing: 24, 46, 80, 84, 85, 206 Squadrons
89th Wing: 20, 23, 92, 101, 211, 218 Squadrons
5th Balloon Wing: 13, 14 and 15 Companies
4th Aircraft Park
4th Air Ammunition Column
4th Reserve Lorry Park
12th Reserve Lorry Park
No. 7 Salvage Section

10th Brigade (attached 5th Army)

80th Wing: 54, 88, 103 Squadrons; 4th Squadron AFC
81st Wing: 2, 21, 42 Squadrons; P Flight
8th Balloon Wing: 3, 11, 20 Companies
10th Aircraft Park
10th Air Ammunition Column
9th Reserve Lorry Park

Appendix 5

FIGHTER COMMAND ORDER OF BATTLE, JULY 1940 (THE BATTLE OF BRITAIN)

Fighter Command Headquarters were at RAF Stanmore.

No. 10 (Fighter) Group

RAF Tangmere 43 Squadron 145 Squadron 601 Squadron Fighter Interception Unit	*RAF Middle Wallop* 238 Squadron 609 Squadron
RAF Pembrey 92 Squadron 213 Squadron	87 Squadron
St Eval 234 Squadron	*Roborough* 247 Squadron
RAF Aston Down 5 Operational Training Unit	*RAF Filton* 6 Anti Aircraft Cooperation Unit
RAF Hawarden 7 Operational Training Unit	*RAF Sutton Bridge* 6 Operational Training Unit

No. 11 (Fighter) Group

RAF Biggin Hill 32 Squadron 610 Squadron	*RAF Hornchurch* 54 Squadron 65 Squadron
RAF Hendon 24 Squadron 257 Squadron	*RAF Kenley* 64 Squadron 615 Squadron
RAF Northolt 1 Squadron 303 Squadron Air Fighting Development Unit	*RAF Manston* 600 Squadron
RAF North Weald 56 Squadron 151 Squadron *RAF Lympne*	*RAF Hawkinge* 79 Squadron

1 Canadian Squadron	*Croydon*	*111 Squadron*	*Croydon*
501 Squadron	*Croydon*	604 Squadron	*Gravesend*
74 Squadron	*Southend*		

No. 12 (Fighter) Group

RAF Coltishall 66 Squadron 242 Squadron	*RAF Debden* 17 Squadron 85 Squadron
RAF Digby 48 Squadron 29 Squadron 511 Squadron	*RAF Duxford* 19 Squadron 264 Squadron 310 Czech Squadron
RAF Wittering 23 Squadron 229 Squadron	*RAF Kirton in Lindsey* 222 Squadron 253 Squadron 266 Squadron
RAF Martlesham 25 Squadron	*RAF Speke* Care & Maintenance Party

No. 13 (Fighter) Group

RAF Acklington 72 Squadron 152 Squadron	*RAF Castletown* 504 Squadron
RAF Catterick 41 Squadron 219 Squadron	*RAF Church Fenton* 73 Squadron
RAF Drem 602 Squadron 605 Squadron 302 Polish Squadron	*RAF Leconfield* 249 Squadron 616 Squadron
RAF Turnhouse 141 Squadron 245 Squadron	*RAF Prestwick*
RAF Usworth 607 Squadron	*RAF Grangemouth* 263 Squadron
RAF Wick 3 Squadron	*RAF Dyce* 603 Squadron
RAF Sunburgh, Shetlands 232 Squadron	

Appendix 6

BOMBER COMMAND ORDER OF BATTLE, MAY 1942

No. 1 (Bomber) Group

RAF Binbrook 12 Squadron 142 Squadron No. 1481 (Gunnery) Flight	*RAF Hemswell* 300 (Polish) Squadron 301 (Polish) Squadron
RAF Elsham Wolds 103 Squadron	*RAF Lindholme* 805 (Polish) Squadron
RAF Holms 1520 (BAT) Flight	*RAF Snaith* 150 Squadron
RAF Breighton 460 Squadron (RAAF)	

No. 2 (Bomber) Group

RAF Horsham St Faith 105 Squadron No. 1508 (BAT) Flight No. 1444 Flight	*RAF Wattisham* 18 Squadron No. 1517 (BAT) Flight
RAF Swanton 88 Squadron	*RAF Watton* 21 Squadron
RAF West Raynham 107 Squadron 114 Squadron No. 1482 (Gunnery) Flight	

No. 3 (Bomber) Group

RAF Feltwell
57 Squadron
75 (NZ) Squadron
No. 1519 (BAT) Flight

RAF Oakington
7 Squadron
101 Squadron

RAF Honington
9 Squadron
No. 1505 (BAT) Flight
No. 1513 (BAT) Flight
No. 1521 (BAT) Flight

RAF Stradishall
109 Squadron
214 Squadron
No. 1483 (Gunnery) Flight

RAF Marham
115 Squadron
218 Squadron
No. 1418 Flight

RAF Tempsford
138 Squadron
161 Squadron

RAF Mildenhall
149 Squadron
419 Squadron RCAF
No. 1503 (BAT) Flight
No. 1429 Flight

RAF Wyton
15 Squadron
156 Squadron
No. 1504 (BAT) Flight

RAF Waterbeach

No. 4 (Bomber) Group

RAF Driffield
158 Squadron
No. 1502 (BAT) Flight
No. 1484 Flight

RAF Middleton St George
76 Squadron
78 Squadron
No. 1516 (BAT) Flight

RAF Linton on Ouse
35 Squadron

RAF Pocklington
405 Squadron RCAF

RAF Leeming
10 Squadron
No. 1512 (BAT) Flight

RAF Topcliffe
102 Squadron

RAF Marston Moor

No. 5 (Bomber) Group

RAF Bottesford 207 Squadron 83 Squadron No. 1518 (BAT) Flight No. 1485 Flight	*RAF Scampton* 49 Squadron
RAF Swinderby 50 Squadron	
RAF Coningsby 97 Squadron 106 Squadron No. 1514 (BAT) Flight	*RAF Waddington* 44 Squadron 420 Squadron No. 1506 (BAT) Flight
RAF Syerston 50 Squadron	*RAF North Luffenham* 408 Squadron (RCAF)
RAF Ossington	*RAF Wing*

No. 8 (Bomber) Group

RAF Chelveston, RAF Molesworth, RAF Thurleigh, RAF Polebrook

No. 91 (Bomber) Group

RAF Abingdon No. 10 OTU No. 1501 (BAT) Flight	*RAF Bassinbourn* No. 11 OTU No. 1446 Flight
RAF Bramcote No. 18 (Polish) OTU	*RAF Chipping Warden* No. 12 OTU
RAF Harwell No. 15 OTU No. 1443 Flight	*RAF Kinloss* No. 19 OTU
RAF Lichfield No. 27 OTU	*RAF Lossiemouth* No. 20 OTU
RAF Moreton in the Marsh No. 21 OTU	*RAF Pershore* No. 23 OTU
RAF Wellesbourne Mountford No. 22 OTU (under direct Air Ministry command)	*RAF Medmenham* Central Interpretation Unit
RAF Benson The King's Flight	

No. 92 (Bomber) Group

RAF Bicester No. 13 OTU No. 1442 Flight	*RAF Cottesmere* No. 14 OTU
RAF Finningley No. 25 OTU No. 1507 (BAT) Flight	*RAF Honeybourne* No. 24 OTU
RAF Upper Heyford No. 16 OTU No. 1511 (BAT) Flight	*RAF Upwood* No. 17 OTU
RAF Wing No. 26 OTU No. 1428 Flight	*RAF Wymeswold* No. 29 OTU

Appendix 7

AIR PAPERS AT THE NATIONAL ARCHIVES WHICH MAY CONTAIN INFORMATION USEFUL TO FAMILY HISTORIANS

AIR 1

Papers used by the Air Historical Branch in writing the official history of the Great War in the Air, including records from the Admiralty, Air Ministry, Ministry of Munitions and War Office. The papers run from *c*.1880 to *c*1922 and are invaluable for research into the First World War.

AIR 2

Policy, case, committee and miscellaneous papers and reports over the whole range of British air administration and related topics from the First World War to the 1970s. These are 'working' files which contain invaluable information on, for example, medal citations.

AIR 3

Airship Log Books 1910–1930. Useful if your ancestor was an airship man.

AIR 4

A small representative selection of flying log books, mainly of Royal Air Force aircrew, but including some log books of Commonwealth and foreign personnel.

AIR 5

Papers of the Air Historical Branch which include some material which plugs gaps in other series for the 1920s, particularly relating to service overseas.

AIR 13

Records of the organization, equipment and operations of the Balloon Command.

AIR 14

Bomber Command papers on operational and technical matters. including many technical reports dealing with aircraft, aircraft losses, armaments, bombing techniques, navigational and photographic aids, and other equipment, operational orders and reports, Air Ministry Directives, the 'Bomber's Baedecker', damage diagrams, Day and Night Bomb Raid Sheets, Interception and Tactics Reports, interpretation reports, orders of battle, raid reports, summaries of Form E reports of bombs dropped on targets in Occupied Europe and War Albums of photographs showing German cities before and after raids. Also included are files on individual operations involving Bomber Command. Invaluable if your relative served in a bomber squadron.

AIR 15
Coastal Command – organization, planning, equipment and operations.

AIR 16
Fighter Command records – organization, defence schemes, trials, training, etc. Invaluable if your ancestor was in a fighter squadron after 1936.

AIR 17
Maintenance Command – organization and activities.

AIR 18
Proceedings of district, general and field general courts martial of officers and other ranks of the Royal Air Force. Some open documents up until the 1960s.

AIR 21
Courts martial registers of charges giving the name and rank of each prisoner, place of trial, nature of the charge and sentence.

AIR 23
Reports, correspondence, etc. on operations of overseas commands (Mediterranean, North Africa, Middle East, India and the Far East) during the Second World War. Includes War Diaries of Air HQ, RAF Iraq, 1923 to 1930.

AIR 24
Daily record of events in each Home Command with appendices illustrating and expanding the record. They include records of the Directorate of the Women's Auxiliary Air Force.

AIR 25
Operations Record Books and appendices for RAF Groups, including some ORBs for Royal Canadian and Indian Air Force Groups which served under RAF command during the Second World War.

AIR 26
Operations Record Books and appendices arranged chronologically under Wings.

AIR 27
Squadron Operations Record Books with appendices including some photographs. There are some ORBs for Dominion and Allied air force squadrons under British command.

AIR 28
Operation Record Books and appendices for RAF stations. It includes some ORBs for RAF Stations which were used by Dominion and Allied air force units.

AIR 29
Operation Record Books and appendices for units of the RAF miscellaneous units, including some ORBs for Dominion air force units operating as part of the RAF during the Second World War.

AIR 30
Submissions to the Sovereign for royal approval of appointments, promotions, awards and regulations, and petitions in court martial cases.

AIR 35
The few surviving records of the British Air Forces in France between September 1939 and May 1940, dealing mainly with administrative and strategic matters.

AIR 36
Records of the Air Component of the North West Expeditionary Force and its operations in Norway (1940).

AIR 37
Records of the Allied Expeditionary Air Force in Europe after D Day and of its principal components the Second Tactical Air Force (RAF), the American Ninth Air Force, USAAF and the Supreme Headquaters Allied Expeditionary Force (Air). Also included are files of, or relating to, 2, 38, 46, 83, 84 and 85 Groups and 310 Squadron, which were controlled by the British organizations above.

AIR 38
Records of the Atlantic Ferry Organization (ATFERO), Ferry Command and Transport Command, including policy files, minutes of committee meetings, orders of battle and route books.

AIR 39
Army Co-operation Command – the organization and operations papers.

AIR 40
Air Intelligence papers on enemy aircraft, weapons, tactics – includes interrogation reports of Allied prisoners of war.

AIR 43
Letters of the Judge Advocate General's Office relating to the preparation of charges for trial by courts martial of Royal Air Force personnel. Largely administrative papers.

AIR 44
Decisions and rulings of the Judge Advocate General relating to courts martial. Largely administrative papers.

AIR 50
Combat reports of squadrons, wings and groups in Fighter, Bomber and Coastal Commands and of Fleet Air Arm squadrons, of various Commonwealth and Allied units based in the United Kingdom including the United States Army Air Force.

AIR 51
Microfilmed papers of the Mediterranean Allied Air Force headquarters dealing with Allied air operations in the Mediterranean during the Second World War and cover the campaigns in Sicily and Italy.

AIR 54
Microfilm copies of ORBs of various operational and administrative units including wings, squadrons and flights; headquarters, depots and air schools, marine and women's units in Africa, the Middle East and Mediterranean.

AIR 71
Letter books of correspondence concerning Air Force Courts Martial.

AIR 76
Service details for officers of the Royal Air Force, mainly for men discharged before 1920. Although the records were created from April 1918 upon the inception of the RAF, they include details of earlier wartime service in the RFC and RNAS.

AIR 78
Index to airmen and airwomen's service records held in AIR 79 giving name and RAF service number only. Numbers run up until *c*.1975. Useful for finding service numbers for First World War servicemen whose records are open.

AIR 79
Service records for airmen of the RAF who served during the First World War – includes previous service in the RFC and RNAS.

AIR 80
Service records for airwomen who served with the Women's Royal Air Force (WRAF) during the First World War.

Appendix 8

USEFUL BOOKS

First World War and earlier

Alfred Gollin, *No longer an Island: Britain and the Wright Brothers* (Heinemann, 1984) puts into perspective the first British attempts at military aviation and the great debate that took place about aerial defence and how to achieve it.

R Dallas Brett, *History of British Aviation 1908–1914* (reprinted by Air Research Publications, 1988) is a valuable compilation from the early editions of *Flight* magazine. It covers the early air races, the first naval and military fliers and the early flying schools, as well as listing the 863 men (and a few women) who qualified as pilots before August 1914. It also has a useful glossary of early aviation terms.

The War in the Air, by Sir Walter Rayleigh (vol. 1) and H A Jones (vols 2–6) are the official history of the air war and give a good picture of the developments and the people as the war progressed.

I McInnes and J V Webb, *A Contemptible Little Flying Corps* (London Stamp Exchange, 1991) lists all the warrant officers, NCOs and men who served in the RFC prior to the outbreak of the First World War. It gives biographical data on many of them, including their career before joining the RFC, what they did during the war and any subsequent career in the RAF. Many of them were commissioned and some rose to quite high rank. There are many photographs.

Trevor Henshaw, *The Sky their Battlefield: Air Fighting and the Complete List of Allied Air Casualties from Enemy Action in the Great War* (Grub Street, 1995) is the most thoroughly researched compilation of RFC and RNAS casualties on a daily basis from TNA files, casualty cards at the RAF Museum, regimental rolls and recollections and log books at the IWM and other museums. This is an excellent starting point if you think that an ancestor was killed in the flying services during the war.

Christopher Shores, Norman Franks and Russell Guest, *Above the Trenches: A Complete Record of the Fighter Aces and Units of the British Empire Air Forces 1915–1920* (Grub Street, 1990). As well as giving a brief history of the air war and of each aviation unit that took part, the book gives brief biographical details of every pilot who was credited with five or more 'kills' and details the date, time and place of their claims. It has many rare photographs of individuals.

Norman Franks, Russell Guest and Gregory Alegi, *Above the War Fronts: A Complete Record of the British Two-Seater Bomber Pilot and Observer Aces and, the British Two-Seater Fighter Observer Aces* (Grub Street, 1997). Companion to the previous volume, this one details bomber and observer aces. Other volumes cover the Allied and German aces.

Airmen Died in the Great War: The Roll of Honour of the British and Commonwealth Air Services of the First World War, compiled by Chris Holson (J B Hayward & Son, 1995). Gives details of RNAS, RFC, RAF, WRAF and Australian Flying Corps casualties, including brief biographies, date and place of death and place of burial. Information is presented both alphabetically and chronologically.

Christopher Cole and E F Cheesman, *The Air Defence of Great Britain 1914–1918* (Putnam, 1984) details the numerous German zeppelin and aeroplane raids on Britain during the war, as well as the efforts made to intercept and stop them. Each raid is looked at individually and there are details of every sortie against them, along with the pilots who took part.

Ray Sturtivant and Gordon Page, *Royal Navy Aircraft: Serials and Units* (Air-Britain (Historians) Ltd, 1992) is an invaluable book that lists, and gives a brief history of, all RNAS and RAF naval aircraft during the war. Movements of aircraft between units and stations are given, along with mentions of pilots known to have flown them. There are brief histories of every aircraft unit and an index of service personnel, as well as maps and plans of every RNAS base in the United Kingdom. There are many photographs.

Brad King, *Royal Naval Air Service 1912–1918* (Hikoki Publications 1997) is a copiously illustrated history of the RNAS by the Imperial War Museum's former film and video archive officer. As well as a history it contains an order of battle for the RNAS on 31 March 1918 and is indexed by personnel, aircraft, units, ships and places.

Dick Cronin, *Royal Navy Shipboard Aircraft Developments 1914–1931* (Air Britain, 1990): amongst a host of useful information, including many lists of aircrew, this gives all known ships that carried seaplanes during the period covered.

Between the wars

Chaz Bowyer, *RAF Operations 1918–1938* (William Kimber, 1988) gives a brief history of the many various 'small wars' in India and the Middle East that the RAF participated in, as well as their operations in Russia and Somaliland immediately after the war. It has a good bibliography of books from the period dealing with the various operations.

John James, *The Paladins: The Story of the RAF up to the Outbreak of World War II* (Futura Publications, 1991). Based on research using the actual figures contained in the RAF list and containing twenty-six tables of facts including numbers of officers, units and aircraft, this book tries to explain the RAF as it actually was, not as the politicians and senior officers would have liked us to believe it was.

T E Lawrence, *The Mint* (Jonathan Cape, 1955) was written by Lawrence after he had served in the ranks of the RAF. It gives an interesting picture of the life of the 'other ranks' between the wars.

Second World War

Francis K Mason, *Battle over Britain* (Aston Publications, 1990) gives a potted history of the lead up to the Second World War from the air perspective, as well as a history of the Battle of Britain. Appendices include a Luftwaffe order of battle, German intelligence appreciations of the RAF, details of ground defences and a comprehensive set of biographies of Fighter Command participants in the battle.

The Fighter Command War Diaries: The Operational History of Fighter Command, Second Tactical Air Force, 100 Group and Air Defence of Great Britain 1939–45, edited by John Foreman (Air Research Publications, 1996–2002). Vols 1, September 1939 to September 1940; 2, September 1940 to December 1941; 3, January 1942 to June 1943; 4, July 1943 to June 1944; 5, July 1944 to May 1945.

John D R Rawlings, *Fighter Squadrons of the R.A.F. and their Aircraft* (Crecy, 1993)

Martin Middlebrook and Chris Everit, *The Bomber Command War Diaries: An Operational Reference Book* (Viking, 1985) details every raid carried out by Bomber Command during the war, though, sadly, individual squadrons are rarely mentioned. It does give a good idea of the sheer scale of Bomber Command's war.

Sir Charles Webster and Noble Frankland, *The Strategic Air Offensive against Germany 1939–1945*, 4 vols (HMSO, 1961).

Philip J R Moyes, *Bomber Squadrons of the R.A.F. and their Aircraft* (Macdonald and Jane's, 1976).

John D R Rawlings, *Coastal, Support and Special Squadrons of the RAF and their Aircraft* (Jane's, 1982).

C Shores and C Williams, Aces High (Grub Street, 1994) gives the definitive details of British and Commonwealth fighter aces in the Second World War.

John Terraine, *The Right of the Line: The RAF in the European War 1939–1945* (Hodder & Stoughton 1985). A comprehensive, but high-level, history of the RAF during the war.

W R Chorley, *Bomber Command Losses of the Second World War*, 5 vols (Midland Counties Publishing, 1992–8) has the brutal statistics of Bomber Command's war.

Air Commodore Graham Pitchfork MBE, *Shot Down and in the Drink: RAF and Commonwealth Aircrews Saved from the Sea 1939–1945* (The National Archives, 2005).

Other useful books

Major A E Marsh, *Flying Marines* (privately published, 1980) is a splendid example of a work published by a genuine enthusiast. It lists all Royal Marines who flew with the RNAS and/or FAA between 1911 and the early 1970s. It contains rare photographs, biographical summaries and excerpts from combat reports and squadron narratives that RM pilots were involved in.

Ray Sturtivant, John Hamlin and James J Halley, *Royal Air Force Flying Training and Support Units* (Air Britain, 1997).

Andrew Boyle, *Trenchard: Man of Vision* (Collins, 1962). Many men contributed to the development of the Royal Flying Corps and Royal Air Force in the early days, but the service is indebted above all to Hugh Trenchard. Though described as 'inarticulate' – hence his nickname 'Boom' – he was not stupid, and it was his vision for the future of the service that preserved it and developed it after the First World War.

James J Halley, *The Squadrons of the Royal Air Force 1918–88* (Air-Britain Publications, 1989) gives potted histories of every RAF squadron, along with brief details of which bases they were at and the aircraft that they flew.

Action Stations (Patrick Stephens Ltd): these excellent books give potted histories of all the various RFC, RNAS and RAF bases throughout the UK and abroad.

They are organized geographically with each book covering a specific part of Britain but historically they cover most of the twentieth century:

Action Stations 1: Wartime Military Airfields of East Anglia, 1939–1945, by Michael J F Bowyer (1979); updated as Action Stations Revisited: The Complete History of Britain's Military Airfields. 1: Eastern England, by Michael J F Bowyer (Crecy Publishing, 2000).

Action Stations 2: Military Airfields of Lincolnshire and the East Midlands, by Bruce Barrymore Halpenny (2nd edn, 1991)

Action Stations 3: Military Airfields of Wales and the North West, by David J Smith (1981)

Action Stations 4: Military Airfields of Yorkshire, by Bruce Barrymore Halpenny (1982)

Action Stations 5: Military Airfields of the South-West, by Chris Ashworth (2nd edn, 1990)

Action Stations 6: Military Airfields of the Cotswolds and the Central Midlands, by Michael J F Bowyer (1983)

Action Stations 7: Military Airfields of Scotland, the North-East and Northern Ireland, by David J Smith (1983)

Action Stations 8: Military Airfields of Greater London, by Bruce Barrymore Halpenny (1984)

Action Stations 9: Military Airfields of the Central South and South-East, by Chris Ashworth (1986)

Action Stations 10: Supplement and Index, compiled and edited by Bruce Quarrie (1987)

Action Stations Overseas, by Sqn Ldr Tony Fairbairn (1991)

Countryside Books produce a series of well researched and action-packed books which covers the airfields of English counties between 1939 and 1945. The history of each airfield is described, with the squadrons and aircraft based at them and the main operations flown. The effects of the war on the daily lives of civilians, and the constant dangers from raids and night bombing are also detailed. Each volume contains approximately 100 black and white photographs. Most English counties are covered, though sometimes they are grouped together in one volume (Devon and Cornwall, or Northumberland and Durham for example).

Max Arthur, *There shall be Wings: Vivid Personal Accounts of the RAF from 1918 to Today* (Hodder & Stoughton, 1993). Concentrating mainly on the Second World War, but with contributions reaching from the foundation of the RAF to the First Gulf War, this is the RAF as seen by the men and women who fought and flew for her.

RAF Yatesbury: The History (privately published, 2004) is edited by Phil Tomaselli (senior) who served at the camp as part of his training for National Service. As well as a history of RAF Yatesbury itself, and its associated camps at RAF Townsend and RAF Cherhill, it contains many amusing tales of

National Service in the 1950s, when Yatesbury was No. 2 Radio School.

Simon Fowler, Peter Elliott, Roy Conyers Nesbit and Christina Goulter, *RAF Records in the PRO: PRO Readers' Guide 8* (PRO Publications, 1994). The first comprehensive (at the time) guide to RAF records at the then Public Record Office (now The National Archives), this still remains a useful guide.

William Maton, *Honour the Air Forces: Honours and Awards to the RAF and Dominion Air Forces during WWII* (Token Books, 2005) is the best specialist book on RAF medals during the Second World War, giving you the squadron and *Gazette* dates for medals awarded to the end of 1945, though it does omit some gazetted during 1946 for war service.

William Spencer, *Medals: The Researcher's Guide* (National Archives Publications, 2006) is the definitive guide to researching medals, written by TNA's military specialist. It contains tips on how to carry out research, as well as lists of possibly useful references and explanations of the various medals, how they were earned and where to find records of them.

N and C Carter, *The Distinguished Flying Cross and How it was Won 1918–1995* (Savannah Publications, 1998) gives unit and *Gazette* information, as well as quoting citations for immediate awards.

I T Tavender, *The DFM Registers for the Second World War* (Savannah Publications, 2004) updates his previous excellent work *The Distinguished Flying Medal: A Record of Courage* (Hayward Books. 1990) and gives unit and *Gazette* information, as well as the AIR 2 references, so you can check the original documents yourself.

Simon Fowler's *Tracing your First World War Ancestors* (Countryside Books, 2003) and *Tracing your Second World War Ancestors* (Countryside Books, 2006) are two excellent small books which are very useful for the genealogical beginner.

Appendix 9

OTHER USEFUL ARCHIVES, COLLECTIONS AND SOURCES

There are numerous local record offices and other archives that might have material relevant to researches into an individual, a squadron or an RAF station. The ones I have detailed below are merely the main ones.

One possible way of identifying useful material in other archives is by using the Access 2 Archives (A2A) system via TNA's website 'Search the Archives' facility. This allows you to search and browse for information about archives in England and Wales in local record offices and libraries, universities, museums and national and specialist institutions, where they are made available to the public.

The Royal Air Force Museum

The RAF Museum at Hendon contains a vast amount of material, some of it of an official nature that has not been deposited at TNA, but the majority being donated by individuals. Highlights include:

- Royal Aero Club (RAC) Pilot's Certificates, complete with photographs of the individual, from the earliest days of aviation up to the 1930s.
- First World War Casualty Cards. The earliest cards are for the RFC only. After 1 April 1918 they include the men who transferred from the RNAS and continue until about 1928. Created when a man was injured, sick or killed, they give details of his injury and how it occurred.
- Microfilm copies of Accident Record Cards giving details of aircraft accidents from 1929 onwards.
- 'Particulars of Non Effective Account' (1938 to 1947) records, giving details of what happened to the effects and final pay of men killed (or committed as insane). There are usually some details of the casualty and next of kin.
- An extensive collection of aircrew log books covering the whole history of the RFC and RAF, mostly donated by ex-servicemen or their families.
- The Air Transport Auxiliary personnel records (available only to next of kin).
- Plans of RAF stations. Most of these are from the 1930s onwards, though a few are earlier.

- Air Ministry Bulletins from the Second World War, giving details of medal citations for RAF servicemen.
- Material from aviation companies, mostly of a technical nature (over 50,000 drawings from Supermarine Ltd alone), though it does have some company board meeting minutes and a list of employees of Sopwith Ltd from 1918.
- Personal papers from some of the famous female aviation pioneers such as Amy Johnson, Jean Batten and Sheila Scott.
- There is much donated material in the form of photograph albums, diaries and personal papers. These range from Lord Trenchard's papers and letters to material provided by ordinary airmen.
- The library collection holds over 13,000 books, 34,000 periodical volumes (including *Flight*, *Popular Flying* and *The Aeroplane* magazines), over 50,000 manuals, 8,500 air diagrams and 6,000 maps.

The Reading Room at RAF Museum London is open for research by appointment Tuesday to Friday, from 10 a.m. to 5 p.m. Researchers wishing to view material should make an appointment by telephone, fax or email *well in advance* of their proposed visit. Please explain the nature of your research so that they can determine how best to help you.

The Museum is at: Royal Air Force Museum London, Grahame Park Way, London NW9 5LL; tel. 020 8205 2266 (general information).

Imperial War Museum

The Imperial War Museum (IWM) was created at the end of the First World War and now covers all aspects of twentieth-century conflict in which Britain's armed forces have taken part. As well as its function of displaying equipment, memorabilia and artwork relating to twentieth-century warfare, the museum holds extensive collections which may be of use to the family history researcher.

The Museum's 'War in the Air' collections relate predominantly to the roles played by British and Commonwealth forces during the world wars. Material comes from both official and private sources, so that the subject can be studied from the command level down to the experiences of individual service men and women.

Collections include Film and Video, Sound Recordings and Oral History, Documents, Artwork, Books and other Documents and Photographs. Over 150,000 of these were listed on the IWM's searchable on-line database in 2006, but this number continues to grow.

The search engine for the collections is complex and requires quite a lot of experimentation to be sure you've got all the possible results. Searching the photographic archive for '804 Squadron' brings up three photographs of the squadron's Hellcats on HMS *Ameer* in 1945. The captions are quite detailed, one reading:

> A large group of men helping to clear a crashed Grumman Hellcat of *804 Squadron*, Fleet Air Arm from the flight deck of the Escort Carrier HMS AMEER whilst others look on. This photograph was taken during the eleven days of action in the course of the successful landings on the islands of Ramree and Cheduba.

It is also possible, with many of the photographs, to view them on-line and to order copies.

The search engine for documents seems particularly baffling but by typing a key word/phrase, such as '115 Squadron' in the Summary box, five references can be obtained, one from the First World War and four from Second World War, including a vivid description of a bombing raid over Duisberg in July 1942.

The Sound Recordings and Oral History archive holds over 21,000 interviews and recordings. Their search engine is particularly friendly.

Imperial War Museum London, Lambeth Road, London SE1 6HZ, UK; general enquiries: mail@iwm.org.uk; tel. 0207 416 5320/5321, fax 0207 416 5374.

The Fleet Air Arm Museum

The museum is situated on the working Fleet Air Arm Station at Yeovilton in Somerset. As well as illustrating the history of British naval aviation, the museum houses an archive dedicated to the collection of information on naval aviation, much of which may be of use to the family historian.

Highlights of their collection include:

- a bound volume listing all the RNAS ratings that transferred to the RAF in 1918 in numerical order;
- enlistment papers, and possibly a statement of service, for RNAS men who didn't transfer to the RAF, or later transferred back to the Royal Navy;
- copies of the service records of RAF and RNAS officers held at TNA under their references AIR 76 and ADM 273;
- a copy of the RAF Muster List for April 1918;
- a collection of FAA Squadron Record Books mainly from the early 1950s onwards;
- the Squadron 'Line Books' which run in parallel with the Squadron Record Books and are a much less formal view of the squadron's activities;
- a large collection of donated photograph albums from the RNAS onwards;
- an extensive range of aviation magazines including *Flight* magazine (1909 to the present), naval aviation magazine *Flightdeck* from the Second World War onwards, *Naval Review* (1948–98);
- some aircraft records and aircraft log cards from the mid-1950s and airframe log cards from the 1950s and 1960s, with a large number of technical manuals for a wide variety of aircraft and balloons;
- Accident Record Cards from 1941 onwards, but these only refer to pilots involved and ignore observers or other crew members.

The archive charges a reasonable research fee (2007) of £15 (this is about standard for type of work being undertaken) for use of their facilities. As always, with any archive, it is best to contact them first to check whether the kind of material they hold will be of any use. They may be able to check their records for you themselves, which will save a visit. The point of contact for any researcher is: Mrs Jan Keohane,

Fleet Air Arm Museum, Box D6, RNAS Yeovilton, Near Ilchester, Somerset BA22 8HT; email research@fleetairarm.com; tel. 01935 840565; fax 01935 842630.

The Museum of Army Flying

The Museum covers the whole history of Army aviation, including the early Royal Engineers, the RFC, the Glider Pilot Regiment, Air Observation Posts and Army Air Corps.

Highlights of their collections include:

- Donated documents and photograph albums relating to the RFC, in particular those squadrons whose main job was supporting the artillery.
- Copies of AOP squadron operation record books from the Second World War, as well as much other donated material.
- Material on all Glider Pilot Regiment major operations. There are copies of the Operation Record Books for the many stations from which gliders operated during the war, including RAF Brize Norton, RAF Tarrant Rushton and RAF Rivenhall.
- From the 1950s and 1960s a wide range of individual and unit diaries, photograph albums, unit histories and some operation record books.

The archive holds over 2,300 photographic negatives from 1911 to the present day, including individuals, groups of officers and men from the RFC to the AAC, groundcrew, individual aircraft and air stations. There is a small collection of donated log books from the First World War to today, which give details of the type of work a pilot was carrying out.

The library and archive are accessible only by prior arrangement. Enquiries can be addressed through the Museum via enquiries@flying-museum.org.uk. Research fees are charged and are currently (2007) £10 per postal enquiry or £10 per personal visit with additional costs of 20p per sheet (plus VAT) for photocopying. Please be aware that, because little information is held on individuals as such, time is required for information to be looked up and collated from the archive's various sources. As ever with any archive you should provide as much information as you have to help the archivist look for your ancestor and to help prevent duplication of results.

The Museum website is at http://www.flying-museum.org.uk/home.html and includes basic details of the collections and facilities as well as opening times and directions. The address to write to is: The Museum of Army Flying, Middle Wallop, Stockbridge, Hampshire SO20 8DY; enquiries@flying-museum.org.uk; tel. 01264 784461.

The Liddle Collection

This was founded in the 1960s to collect and preserve first-hand individual experiences of the First World War. The archive includes original letters and diaries, official and personal papers, photographs, newspapers and artwork, as well as written and tape-recorded recollections.

Material is arranged within thirty sections, according to nature of service or geographical area, and these are listed in the summary guide, which also contains subject indexes to uncatalogued material where appropriate. The archive can be searched on-line at: http://www.leeds.ac.uk/library/spcoll/liddle/index.htm.

There are 328 RFC references, 162 RNAS references and 383 for the RAF (though of course many overlap with the RFC and RNAS).

By a curious coincidence it holds much information on Lieutenant C P O Bartlett, RNAS, who featured in 'Other officers' records'. The index to his material reads:

> Bartlett, Charles Philip Oldfeld. Lieutenant, RNAS (Royal Naval Air Service) and Squadron Leader, RAF (Royal Air Force). During the First World War he was active on bombing raids in Flanders and over the Somme on the Western Front. He was awarded the DSC in 1917, and a bar to the DSC in Mar 1918 for his activities on the Somme
>
> 4 volumes of photocopied diaries (Sep 1916-Mar 1918); 5 Raid Orders (Mar 1918), with photocopied duplicates; Contemporary typescript account of an aerial engagement over Flanders (Mar 1918); Photocopied Squadron Magazine 'Dope: Unofficial Organ of No. 5 Wing' (Apr 1917); Envelope containing newspaper cuttings relating to Bartlett's honours (1917–1918); 46 photographs, most with captions (1916–1918) and 4 negatives; War Office Map, NW Europe (1915); 4 RNAS Christmas Cards (1916–1917); 2 RNAS printed cartoons (1916–1917); 12 photocopied RNAS cartoons 'The Education of the Quirk' (nd); Manuscript copy of a poem [by R F Slade?] (1917); Typed transcript of an interview recorded with Peter Liddle (July 1976).

The collection was originally established to gather material relating to the First World War, but has been collecting Second World War material for some time. There are 123 collections relating to various aspects of the RAF. A random reference illustrates the nature of the material held:

> Clement, Edwin Leslie (d 1945). Leading Airman, No. 84 Squadron, RAF (Royal Air Force), serving in Egypt, Greece, Iraq, and India. Captured by the Japanese, Feb–Mar 1942, and held POW (Prisoner of War) in Kuching Camp, Sarawak, Borneo. Died in captivity, 5 Sep 1945.
>
> Original letters and postcards to and from EL Clement; Letters of condolence; Various papers; Photographs.

If you are interested in looking at any of the material you have identified you will need to contact the archive in advance to make an appointment. If there is not much material then it may be possible (for a charge) for the relevant material to be copied and posted to you.

Contact in the first place should be made with Richard Davies, Keeper of the Collection, at r.d.davies@leeds.ac.uk. I have always found Richard and his small staff extremely friendly and helpful.

The Air Historical Branch

The Air Historical Branch (RAF) is a small part of the UK Ministry of Defence which seeks to maintain and preserve the historical memory of the RAF and to develop and encourage an informed understanding of RAF and air-power history by providing accurate and timely advice to Ministers, the RAF, other government departments and the general public. The branch is not a public record depository, but maintains a substantial archive of classified policy and operational documents, which are normally declassified after thirty years and transferred to the The National Archives. The branch also holds an index of RAF casualties from 1939 onwards, aircraft accident record cards dating from the inter-war years, and a photographic archive.

Air Historical Branch (RAF), Building 266, RAF Bentley Priory, Stanmore, Middlesex HA7 3HH.

Existing squadrons

Following on from the Air Historical Branch, a frequently overlooked source of information is the records of the current RAF squadrons. These squadrons have inherited, through the AHB, material donated by former members and maintain strong and cordial links with their veterans. I fondly recall spending an evening in the Mess of 47 Squadron at RAF Lyneham in the mid-1990s looking over their photograph albums and memoirs of former squadron members of all ranks. Every squadron has a senior NCO or junior officer whose job it is to keep up the records and maintain links. A list of current squadrons is available on the official RAF website. Please remember that these are operational units and, keen as they may be to assist, they have responsibilities and duties of far greater importance than dealing with family historians. Keep your requests short, to the point and polite, and don't expect an immediate reply! A reply paid envelope is also a courtesy.

Appendix 10

MAGAZINES AND PERIODICALS

Cross and Cockade International is the journal of the First World War Aviation Historical Society and is an invaluable source of information on all aspects of early military and naval aviation. Membership of the society costs £25 (in 2006) and entitles members to the four copies of the journal produced annually. Articles cover individuals, aircraft, units, weapons, organization, camouflage, training, technology and actual air battles. There are regular readers' queries and sources and pointers for further research. They can be contacted initially through their website http://www.crossandcockade.com. The American version is *Over the Front,* a quarterly magazine produced by the League of First World War Aviation Historians.

The Western Front Association exists to further interest in, and remembrance of, the First World War through regular local meetings and through its two journals *Stand To!* and *The Bulletin*. The collective expertise of its membership is vast and extends well beyond the trenches of France and Belgium into all aspects of the war. Their website is at http://www.westernfront.co.uk and this will give you details of your local branch, where and when they meet and who their speakers will be. Thoroughly recommended for anyone with an interest in the First World War.

The Gallipoli Association exists to keep alive the memory of the Gallipoli campaign of 1915 through its newsletter *The Gallipolian*. Their website is at http://www.gallipoli-association.org. Though only a few RNAS aircraft (under the command of the ubiquitous Samson!) took part in the campaign, very many men who served there with the Army and Navy went on to transfer to the flying services.

Fly Past is a monthly magazine that features articles covering the whole history of aviation (not just British). Its website is at http://www.flypast.com.

Local newspapers

The local press is always eager for information about local people because, if nothing else, they have relatives who will want to buy a copy of anything which relates to their family. Your local library will have copies. In the event that you need ones from further afield then the British Library Newspaper Library at Colindale should hold copies.

Appendix 11

WEBSITES

There are an increasing number of websites devoted to specific aspects of the Second World War, including individual units, ships, RAF squadrons and bases. Use your computer's internet search facility to identify sites that you think might be relevant. Please be aware that, while many sites are of a high quality, some are not.

The following is a list of sites that you may find useful to begin with and many link to further sites you may find interesting.

National Archives (formerly Public Records Office), Kew

http://www.pro.gov.uk

Many of the records you are likely to need are held at Kew. Make yourself familiar with this site and with the search engine for the archive and use the download facility to look at their extensive range of leaflets which explain the types of records and where they can be found. This site is well worth your time in advance of any visit to Kew, as familiarity with it will save a lot of precious time during your visit.

The National Archives produce a small series of guides to air services records, which are available on-line so you can download them and read them at your leisure.

http://www.nationalarchives.gov.uk/familyhistory/guide/airforce/default.htm

This guide is specifically designed for the family historian and gives basic details of where to find records on airmen and non-commissioned officers, officers, medals, operational records, courts martial and the Women's Royal Air Force.

http://www.nationalarchives.gov.uk/catalogue/RdLeaflet.asp?sLeafletID=60

This gives a detailed set of sources for First World War servicemen, servicewomen and officers, operational records, courts martial and medals. One point to mention is that the guide does suggest that service records for men from the RFC who did not transfer to the RAF should be looked for in the Army records. This does not seem to be the case – look for them in AIR 79 first and only go to the Army records if you can't find them there.

http://www.nationalarchives.gov.uk/catalogue/RdLeaflet.asp?sLeafletID=61

This is a more detailed guide to RAF service records for the Second World War, including some useful hints on investigating crashed aircraft and a handy bibliography.

http://www.nationalarchives.gov.uk/catalogue/RdLeaflet.asp?sLeafletID=155

This gives details of how to find operational records for the various air services, including the Glider Pilot Regiment and Air Observation Posts for the army and the Fleet Air Arm.

http://www.nationalarchives.gov.uk/catalogue/RdLeaflet.asp?sLeafletID=36

Strictly speaking, this guide is concerned with technical and research records which are unlikely to mention individuals unless they were involved on a specific project or worked at one of the RAF's associated organizations such as the Royal Aircraft Establishment or the Ministry of Aircraft Production.

The Ministry of Defence Website

This is at www.mod.uk and it links to the Veterans Agency site www.veteransagency.mod.uk where relevant details can be found for tracing the records of Army, Navy and Air Force service personnel.

Commonwealth War Graves Commission

http://www.cwgc.org/cwgcinternet

This site will enable you to find details of servicemen and civilians who were killed or died during the war.

RAF Museum at Hendon

www.rafmuseum.org.uk

This provides useful links to various RAF Association websites.

The RAF official website

http://www.raf.mod.uk/rafhome.html

This provides a comprehensive history of the service through its link with the RAF Historical Branch.

The Imperial War Museum

www.iwm.org.uk

This has a somewhat daunting (at first) search engine for their extensive collections of material and some useful links for family historians.

WWII Experience Centre at Leeds

http://www.war-experience.org/index.html

Invaluable because it contains links to dozens of other sites, many belonging to associations such as the Dunkirk Veterans Association, Glider Pilot Regiment Association and the RAF Ex-POW Association. Well worth a browse.

The Fleet Air Arm Archive 1939–1945

www.fleetairarmarchive.net

Please note this is NOT the official site of the Fleet Air Arm Museum. This gives many details of FAA units and actions during the Second World War.

Bomber Command

www.bomber-command.info

This is a private site giving much interesting and useful information on Bomber Command's war.

Second World War RAF

http://www.worldwar2exraf.co.uk

This is a website aimed primarily at those seeking to be reunited with old friends or looking for information on someone who served with the Royal Air Force during the Second World War.

Royal Canadian Air Force

http://www.airforce.ca/index2.php3?page=honours

This part of the website lists honours and awards from the First World War, members of the RCAF in the Second World War and Canadians who served in the RAF during the Second World War.

Appendix 12

RAF ABBREVIATIONS

As a technical service the RAF is particularly fond of abbreviations and acronyms and a service record can appear to be little more than a confusing series of dates, letters and numbers. These can be interpreted and turned into something meaningful, but it may take time. It is always worth it to get the fullest picture of your ancestor's service and once you've worked out what the various acronyms mean you can go looking for further records from the units they were attached to, to find out where they were and what they were doing.

There are some relatively common abbreviations, which may make interpretation easier, but they are by no means guaranteed. F frequently refers to 'Flight', the smallest working independent RAF unit, consisting of three or four aircraft. U is usually 'Unit'; S is frequently 'School', as in AFS (Advanced Flying School), though can also mean Squadron, as in AFS (Auxiliary Fighter Squadron). T is frequently 'Training' as in OTU (Operational Training Unit) but also can mean Technical as in 'Sch of TT' (School of Technical Training). It almost seems possible that you can take any random selection of words from the following list and at some time or another there would have been a unit with that name!

One way of breaking the 'code' is to refer to the RAF List or Confidential List for the period covered by the individual's posting. Though this can be a bit of a haul it should help you identify the unit they were attached to, as well as where they were posted. A most useful guide has been compiled on-line by a group of Second World War RAF veterans drawn from their own experiences and from requests they've received for help over the years. The web address is http://www.worldwar2exraf.co.uk/acronyms%20A.htm. The site lists over a thousand acronyms and tries, where there might be more than one (as in Reception Units), to identify the site of the unit by its number. Many acronyms duplicate over the years as units are broken up and then other units, with a similar name and so identical acronym, are created later. Many units were duplicated by overseas commands so it is worth knowing that these were usually identified further by referring to their Command area such as (ME) Middle East or (I) India, as in AHQ(I) (Air Headquarters (India)). Sometimes it is possible to combine two individual acronyms to work out what another stands for, such as ACF (Air Command Far East) and AWF to produce ACFAWF (Air Command Far East All Weather Flight).

A much abridged list of the more basic abbreviations is set out below but I must stress that it covers only a fraction of the possible acronyms you may encounter. Rank abbreviations are included in Appendix 1.

AAC	Army Air Corps
AAC	Air Ammunition Column (WW1)
AAEE	Aeroplane and Armament Experimental Establishment
AAF	Auxiliary Air Force
AAP	Aircraft Acceptance Park (during WW1)/ Army Aircraft Park
AFEE	Airborne Forces Experimental Establishment
A/C	Aircraft
ACC	Army Co-operation Command
ACDW	Air Crew Disposal Wing
ACHU	Aircrew Holding Unit
ACMB	Aviation Candidates Medical Board
ACRC	Air Crew Reception Centre
ACS	Airship Construction Station/Service (WW1)
ACSB	Air Crew Selection Board
AD	Aircraft Depot
ADGB	Air Defence of Great Britain
ADRU	Air Despatch and Reception Unit
ADU	Aircraft Delivery Unit
AED	Aircraft Equipment Depot
AF	Advanced Flying/Air Fighting
AFC	Australian Flying Corps (WW1)
AFE	Airborne Forces Establishment
AFS	Auxiliary Fighter Squadron
AGBS	Air Gunnery and Bombing School
AGRS	Advanced Ground Radio School
A&GS	Armament and Gunnery School
A&IC Sch	Artillery and Infantry Cooperation School
Air Min /AM	Air Ministry
ALG	Advance Landing Ground
ALO	Air Liaison Officer
ANBS	Air Navigation and Bombing School
AOC	Air Officer Commanding
AO	Air Observer
AOP	Air Observation Post
AP	Aircraft
ARC	Aircrew Recruiting Centre/Aircrew Reception Centre
ARD/F	Aircraft Repair Depot/Flight
ARP/S/U	Aircraft/Aeroplane Repair Park/Section/Unit
A/S	Anti Submarine
ASC	Aircrew Selection Centre

ASD	Aeroplane Supply Depot
ASP	Air Stores Park
ASR	Air Sea Rescue
AS&RU	Aircraft Salvage and Repair Unit
ASS	Air Signals School
ASU	Aircraft Storage Unit
ATA	Air Transport Auxiliary
ATC	Air Traffic Control/Air Training Corps
ATFERO	Atlantic Ferry Organisation
ATS	Auxiliary Territorial Service (Women)
ATS	Armament Training Station/Air Training Squadron
AWF	All Weather Flight
AWOL	Absent Without Leave
B&C	Barrack and Clothing
BANS	Basic Air Navigation School
BBOC	Brought Back on Charge (applies to repaired aircraft)
BC	Bomber Command
BCAS	Bomber Command Armament School
BCATP	British Commonwealth Air Training Plan
BCBS	Bomber Command Bombing School
BCMC	Bomber Command Modification Centre
BCo	Balloon Company (WW1)
Bde	Brigade
BDU	Bomb Disposal Unit
BER	Beyond Economical Repair
B&GS	Bombing and Gunnery School
BMH	British Military Hospital
BRD	Base Repair Depot
BSDU	Bomber Support Development Unit
BTU	Bombing Trials Unit
BTW	Balloon Training Wing (WW1)/Boys Training Wing (WW1)
BU	Broken Up
(C)	Coastal
CACF	Coast Artillery Co-operation Flight
CAEU	Casualty Air Evacuation Unit
CF/Unit	Camouflage Flight/Unit
C&M	Care and Maintenance
Cam-ship	Catapult-armed Merchantship
CARD	Central Aircraft Repair Depot
Casevac	Casualty Evacuation

CAW	College of Air Ware fare
CBCF/S	Coastal Battery Cooperation Flight/School
CC	Confined to camp
CCFATU	Coastal Command Fighter Affiliation and Training Unit
CCS/H	Casualty Clearing Station/Hospital (WW1)
CD	Clothing Depot (WW1)
Cdt Brig	Cadet Brigade (WW1)
CF	Communication Flight/Conversion Flight
(C)FPP	Civilian Ferry Pilots Pool
CF(S)	Communications Flight (Squadron)
CFS	Central Flying School
CGIS	Central Gliding Instructors School
CGS	Central Gunnery School/Central Gliding School
Cmd	Command
CMU	Civilian Maintenance Unit
CNS	Central Navigation School
CO	Commanding Officer (can also be OC)
Comm(s)	Communication(s)
Conv	Conversion
CPE	Central Photographic Establishment
CRE	Central Reconnaissance Establishment
CRO	Civilian Repair Organization
CRP	Civilian Repair Party
CS	Communication Squadron
CSE	Central Signals Establishment
CSF/S	Communications and Support Flight/Squadron
Cse	Course
CTS	Combat Training School
CU	Conversion Unit
CW	Communication Wing
CWg	Cadet Wing (WW1)
DAF	Desert Air Force
DBF	Destroyed by Fire (applies to aircraft)
DBR	Damaged beyond repair (applies to aircraft)
Del/Dly	Delivery
Det/Dett	Detachment (detachment to another unit or Squadron)
DI	Daily Inspection
Disb	Disbanded
DS	Depot Squadron (WW1)
DU	Development Unit

E/A	Enemy Aircraft
EAAS/NS	Empire Air Armament School/ Navigation School
EAOS	Elementary Air Observer School
EATS	Empire Air Training Scheme
EDD	Equipment Disposal Unit
EFTS	Elementary Flying Training School
EGS	Elementary Gliding School
ELG	Emergency Landing Ground
EPD	Equipment and Personnel Depot
ERS	Empire Radio School/Engine Repair Section
ETB	Eastern Training Brigade (WW1)
ETPS	Empire Test Pilots School
EOWS	Elementary Observer Wireless School
EU	Equipment Unit
Evd	Evaded Capture
FAA	Fleet Air Arm
FAGS	Fleet Aerial Gunners School
FBDF/SU	Flying Boat Development Flight/Servicing Unit
FC	Ferry Command/Fighter Command/Flying Control
FCCS/F	Fighter Command Communication Squadron/Flight
FCITF/S	Fighter Command Instrument Training Flight/.Squadron
FCPU	Ferry Command Preparation Unit
FCRS	Fighter Command Radar School
FCTU	Fighter Command Trials Unit
F/E	Flight Engineer
FE	Far East
FECEF	Far East Casualty Evacuation Unit
FEE	Fighter Experimental Establishment
FEFBW	Far East Flying Boat Wing
FFI	Free From Infection
FIS	Flying Instructors School
FLS	Fighter Leaders School
Flt	Flight
FPP	Ferry Pilots Pool
FR	Fighter-reconnaissance
FRD	Forward Repair Depot
FRU	Field Repair Unit
FS	Fighting School
FSAF&G	Fleet School of Aerial Fighting and Gunnery School
FSS	Ferry Support Squadron/Flying Selection Squadron

FTC	Flying Training Command
FTR	Failed to Return
FU	Ferry Unit
G	Glider/Gliding
GATU	Ground Attack Training Unit
GC	Gliding Centre
GD	General Duties
GDC	Group Disbandment Centre
GED	Ground Equipment Depot
GES	Glider Exercise Squadron
GG	Ground Gunnery
GI	Ground Instructional
GIF	Glider Instructors Flight
Gp	Group
GPUTF	Glider Pick-up Training Unit
GPR	Glider Pilot Regiment
GRSS	Ground Radio Servicing Squadron
GRU/F	General Reconnaissance Unit/Flight
GSE	Glider Servicing Echelon
GTF	Gunnery Training Flight
GTPS	Glider & Tug Pilots School
GTS	Glider Training Squadron/Glider Training School
HAD	Home Aircraft Depot
(HB)	Heavy Bomber
HCCS	Home Command Communications Squadron
HCF	Helicopter Communications Flight/Home Communications Flight
HD	Home Defence (WW1)
HDU	Helicopter Development Unit
HE	Home Establishment
HFF	Heavy Freight Flight
HFU	Home Ferry Unit
HG	Heavy Glider
HOCF	Helicopter Operational Conversion Flight
Hosp	Hospital
HS Flt	High Speed Flight
HT	Heavy Transport
(I)	India
IAAD	Inland Area Aircraft Depot
IAF	Indian Air Force
i/c	in charge

IFDU	Intensive Flying Development Unit
IFF	Identification Friend or Foe
IFTS	Initial Flying Training School
Int	Interned
I/O	Intelligence Officer
IR	Immediate Reserve
ISF	Internal Security Flight
IT	Initial Training
JAPIC	Joint Air Photographic Interpretation Centre
JARIC	Joint Air Reconnaissance Intelligence Centre
JCU	Jet Conversion Unit
JEHU	Joint Experimental Helicopter Unit
JHDU	Joint Helicopter Development Unit
JTF	Jet Training Flight
JSPI	Joint School of Photographic Interpretation
JSSC	Joint Services Staff College
JSTU	Joint Services Trials Unit
KB	Kite Balloon (WW1)
KBS	Kite Balloon Section (WW1)
KBT	Kite Balloon Training (WW1)
KD	Khaki drill
KF	King's Flight
KRs	King's Regulations
LAAGS	Light Anti-Aircraft Gunnery School
LAS	Light Aircraft School
LG	Landing Ground
LLF	Light Liaison Flight
LRFU	Long Range Ferry Unit
LRDF	Long Range Development Flight
LRWRE	Long Range Weapons Research Establishment
MA	Marine Aircraft
MAC	Mediterranean Air Command
MAD	Marine Acceptance Depot
MAP	Ministry of Aircraft Production
MASR&CF	Malta Air Sea Rescue & Communications Flight
MATAF	Mediterranean Allied Tactical Air Forces
MC	Maintenance Command
MCA	Ministry of Civil Aviation
M&D	Medicine and Duties
ME	Middle East

MED	Medical Equipment Depot
Med Dist	Mediterranean District
MEP&AP	Middle East Pilot & Aircrew Pool
Met Res	Meteorological Research
MFPU	Mobile Field Photographic Unit
MGSP	Mobile Glider Servicing Party
MO	Medical Officer
MOA	Ministry of Aviation
MORU	Mobile Operations Room Unit
MOS	Marine Observation School
MF	Meteorological Flight
MRS	Maritime Recognisance School
MRU	Medical Rehabilitation Unit
MT	Motor Transport
MTDpt	Marine Training Depot
MTE	Medical Training Establishment
MTLRU	Motor Transport Light Repair Unit
MU	Maintenance Unit
Nav	Navigator
Navex	Navigation Exercise
NCU	Night Conversion Unit
N/E	Non Effective
NFDS/W/U	Night Fighter Development Squadron/Wing/Unit
NFF	Night Fighter Flight
NFT	Night Flying Test
NFTS	Night Fighter Training Squadron
NICF	Night Conversion Flight
NTS	Night Training Squadron
NVTS	Night Vision Training School
(O)	Observer
OADF/U	Overseas Aircraft Delivery Flight/Unit
OANS	Observers Air Navigation School
OAPU	Overseas Aircraft Preparation Unit
OATS	Officers Advanced Training School
Obs Sch/OBS	Observer School
OCU	Operational Conversion Unit
OCW	Officer Cadet Wing (WW1)
OEU	Operational Evaluation Unit
OFU	Overseas Ferry Unit
OP	Observation Post (Army Air Corps unit)

Opl	Operational
ORTU	Operational and Refresher Training Unit
OS	Observers School
OSR&AP	Observers School of Reconnaissance and Aerial Gunnery
OTU	Operational Training Unit
OTTW	Officer's Technical Training Wing
PACT	Pre Aircrew Training
(P)	Pilot
Pal Bde	Palestine Brigade (WW1)
PD	Packing Depot
PDRC	Personnel Dispatch and Reception Centre
PDU	Personnel Dispersal Unit
PDU	Photographic Development Unit
PF	Practice Flight
PFF	Pathfinder Force
PGTS	Parachute and Glider Training School
PHU	Personnel Holding Unit
PIU	Photographic Intelligence Unit
POW	Prisoner of War
PPF	Parachute Practice Flight
PPP	Pupils Pilots Pool
PR	Photographic Reconnaissance
PRC	Personnel Reception Centre
PRDU/E	Photographic Reconnaissance Development Unit/Establishment
PRFU/S	Pilots Refresher Flying Unit/School
PRU	Photographic Reconnaissance Unit/Pilot Replacement Unit
PRU	Personnel Reception Unit
PS	Parachute School
PTC	Parachute Training Centre/Personnel Transit Camp
PTF	Parachute Test Flight
PTU	Parachute Test Unit/Parachute Training Unit
PTURP	Pilots Training Unit and Reinforcement Pool
QF	Queen's Flight
QFI	Qualified Flying Instructor
QRs	Queen's Regulations
(R)	Reserve
RAAF	Royal Australian Air Force / Royal Auxiliary Air Force
RAE	Royal Aircraft/Aerospace Establishment
RAFC	Royal Air Force College
RAFFC	Royal Air Force Flying College

RAFG	Royal Air Force Germany
RAFR	Royal Air Force Regiment
RAFSC	Royal Air Force Staff College
RAFVR	Royal Air Force Volunteer Reserve
RAP	Reserve Aircraft Pool
RAS	Reserve Aeroplane Squadron
RC	Recruiting Centre
RC	Reserve Command
RCAF	Royal Canadian Air Force
RD	Recruits Depot or Reserve Depot (WW1)
RDUK	Repairable at Depot in UK
Recce	Reconnaissance
RFC	Royal Flying Corps
RFF/U	Refresher Flying Flight/Unit
RFS	Reserve Flying School
R/G	Rear Gunner
RLP	Reserve Lorry Park
RMU	Radio Maintenance Unit
RNAS	Royal Naval Air Station/Service
RNEFTS	Royal Navy Elementary Flying Training School
ROC	Royal Observer Corps
ROS	Repairable on Site
RRF(U)	Radio Reconnaissance Flight (Unit)
RS	Radio School/Reserve School/Reserve Squadron
R&SS	Repair and Salvage Section
RSU	Repair and Servicing Unit/Repair and Salvage Unit
R/T	Radio Telegraphy/Radio Telephony
RTP	Recruit Training Pool
RTU	Return To Unit
RU	Reception Unit (of people)/Repair Unit (of aircraft)
RWE	Radio Warfare Establishment
S/Sch	School
SA	South African
SAR	Search and Rescue
SARD	Southern Aircraft Repair Depot
SAS	Servicing Aircraft Section
SC	Salvage Centre
SCR	School of Control and Reporting
SCS	Special Communication Squadron
SD	Special Duty/Special Duties/Stores Depot

SDP	Stores Distributing Park
SDU	Signals Development Unit
S/E	Single Engined
SEAC	South East Asia Command
SFTS	Service Flying Training School
SGR	School of General Reconnaissance
SHAEF	Supreme Headquarters Allied Expeditionary Force
Sigs	Signals
SIU	Signals Intelligence Unit
S/L	Searchlights
SL	Sick leave
SLG	Satellite Landing Ground
SMR	School of Maritime Reconnaissance
SNBD	School of Aerial Navigation and Bomb Dropping
SNC	School of Naval Cooperation
S of M A	School of Military Aeronautics (WW1)
SOC	Struck off Charge
SOF	Special Operations Flight
SoFP	School of Fighter Plotting
SofP	School of Photography
SoN&BD	School of Navigation and Bomb Dropping
SoNC&AN	School of Naval Cooperation and Aerial Navigation
SoSF	School of Special Flying
SPTU	Staff Pilots Training Unit
Sq/Sqn/Sqd	Squadron
SR	Strategic Reconnaissance
SRCU	Short Range Conversion Unit
SROs	Station Routine Orders
SRTS	Short Range Transport Squadron
SRU/W	Strategic Reconnaissance Unit/Wing
SS	Signals School/Salvage Section
SSF	Special Service Flight/Special Survey Flight
Stn	Station
STS	Seaplane Training Squadron/Special Transport Flight
S&TT	Station and Target Towing
SU	Support Unit/Servicing Unit
Supy	Supernumary
SWTS	Supplementary Wireless Telegraphy School
T	Training/Transmitter
TAF	Tactical Air Force

TAG	Telegraphist Air Gunner (Fleet Air Arm)
TA	Transport Aircraft
TB	Torpedo Bomber
TC	Training Command/ Transport Command
TCAEU	Transport Command Aircrew Examining Unit
TCICU	Transport Command Initial Conversion Unit
TD	Tent Detachment (WW1)
TEE	Test and Evaluation Establishment
TF	Training Flight
TFPP	Temporary Ferry Pilots School/Training Ferry Pilots Pool
TFP	Training Ferry Pool
TFU	Telecommunications Flying Unit
TPS	Test Pilots School
TPO	Teleprinter Operator
T/R	Transmitter/Receiver
Trg	Training
TS	Training School/Training Squadron
TSCU	Transport Support Conversion Unit
TSR	Torpedo Spotter reconnaissance
TT	Target Towing
T.T.	Trade Training
TTC	Technical Training Command
TSTS	Trade Selection Test Section
TTS	Torpedo Training School/Squadron
TTU	Torpedo Training Unit/ Target Towing Unit
TU	Training Unit
u/I	unidentified
us	unserviceable
USAAC/F	United States Army Air Corps/Force
UAS	University Air Squadron
UED	Universal Equipment Depot
W or Wg	Wing
W&OS	Wireless and Observers School
WAAF	Women's Auxiliary Air Force
WEE	Wireless Experimental Establishment
WIDU	Wireless Intelligence Development Unit
W/Op	Wireless Operator
WOp/AG	Wireless Operator/Air Gunner
WRAF	Women's Royal Air Force
WRAFVR (T)	Women's Royal Air Force Volunteer Reserve (Training branch)

WS	Wireless School
WT	Wireless Telegraphy
WTt	Wireless Telephony

INDEX

Aden (RAF in) 86
Admiralty 63, 121, 122, 123, 135
Air Board (formation of) 65
Air Observation Posts 123–124
  Gallantry recommendations 125
  Operation Record Books 123–124
  Post WW2 128
  Service Records 125
  War Diaries 125
  (*see also Glider Pilots*)
Air Transport Auxiliary 125–126
Annesley, Flying Officer W R B, 22
Armoured Car Squadron (RNAS) 60–62
Armoured Car Squadrons (RAF) 118–119
Army Pilots (post WW1) 123–125
Ashmore, Brig General E B 38, 40, 41

Balloons (RFC) 32 33
Barkway, Sgt pilot Geoffrey 125
Bartlett, Sqd Cdr CPO 54–55, 56
Battle of Britain 89–95
Bebbington, Flt Sgt F 151
Bechuanaland 8, 13, 139
Berlin Airlift 132–133
Berlyn, Lt R C 73–74
Blampied, 2nd Lt Guy, 70
Blazey, AM1 G W 57
Boer War 8, 12, 13, 139
Bomber Command 95–99
Bone, Group Capt R J 1, 2, 3, 16, 80, 85, 89, 156–7
Bremner Sqdn Cdr FDH 52–53
Bromley, Flt Sgt J E 107
Burness, Samuel 10–12
Busteed, Wng Cdr H R 87

Capper, John E 8, 13
Casualties 152–153
Central Flying School 14, 17
Chanak incident 86
China 12, 13, 123, 139, 140
Clarabut, Lt PG 73
Coastal Command 99–102
Cody 'Colonel' S F 14, 15, 16
Cologne (Thousand Bomber Raid) 96–99
Combat Reports 93, 94, 98,
Commands (RAF) 88
Cooke, Flt Sgt Ronald 112–116

Dew, AM A G 63–64
Douglas-Pennant, Lady Violet 67
Dowding, ACH Sir Hugh 41, 42, 89, 91
Drewett, Chief Mechanic Ernest 25
Drigh Road (Karachi) 79–82

Evans, Flying Officer D L 150

Fairclough, PMO J G 57
Fighter Command 88, 89–95
Fleet Air Arm (FAA)
  Combat reports 123
  During WW2 119–123
  Flights 87
  Naval Officers records 87
  ORBs 121
  Part of RAF 86–87
  Post WW2 128, 129
  Returns to Navy 87
  Service records 123
  Squadrons 87, 120–121
  Suez 135

Glaser, Pilot Officer 93
Glider Pilot Regiment
  Operation Record Books 123–124
  Post WW2 128
  Service Records 125
  War Diaries 125
Guy, Lt Christopher Guy 46–49

Hainsworth, 2nd Lt G 27, 28
Harris, AM Sir Arthur 'Bomber' 77, 96
Hast, Sub Lt E F 52
Heath, 2AC Reginald 109–111
Hobson, 2AC Ernest .23, 24
Horton, AM1 John 144
Horsley, Pilot Officer R M 150

India (RAF in) 78–85, 146
Intelligence School (RAF?) 105–106
Iraq (RAF in) 76–78, 146
Ivelaw-Chapman, ACM Sir R 82

Jarvis, AC1 149
Jellicoe (Armoured Train) 62
Jones, AC1 Vic 81

Keith, Sqd Leader CH 78
Kenya (RAF in) 134
Korean War 133

Locker-Lampson, Cdr Oliver 60
Log Books 112–116
London Gazette 140–142
Longmore, ACM Sir Arthur 16, 154

Mackenzie, Lt J H L (RE) 9, 13
McLean, Driver, J R 25
Malaya (RAF in) 134
Medals 10, 71, 124, 139–152
  London Gazette 140–142
  WW1 Campaign 142–143
  WW1 Gallantry 142–143
  Between the Wars 145–147
  WW2 145, 148–151
  WW2 Campaign 147–148
  WW2 Gallantry 148–151
  Post WW2 151–152
Mesopotamia 66, 72

National Service 129–131
Nicholas, Observer Officer R 156–157
Nuclear Weapons 128

Operations Record Books (ORBs) 74, 77, 79, 82, 84, 87, 92–93, 97, 98, 105, 106, 116, 117, 119, 120, 123, 125, 127, 128, 129
Packe, 2nd Lt E J 27
Palestine (RAF in) 132
Parry, Lt Cdr P RNVR 154
Pascoe, 2nd Lt A 27
Patterson, J W 19, 20, 21, 22, 90, 91
Pickthall, Lt RNVR 52, 54
Pink, Air Commodore Richard 83

Rhine Army (RAF with) 68–69
Richardson, Corporal F J 25
Rodgman Sergeant A G 25
Rogers, Lt W P RNVR 52, 54
Ross, Flying Officer J 83
Royal Air Force
  Commands 88
  Expansion 88
  Formation 65
  Nurses 117–118
  Regiment 118–119
  Reserve 73
Royal Air Force (Stations)
  Calshot 156–157
  Cranwell 57
  Dover 53
  Drigh Road 79–82
  Dunkirk 56, 57, 59, 63
  Eastchurch 16, 52
  Elsham Wolds 97
  Exeter 116
  Farnborough 15
  Great Yarmouth 37, 58, 59, 63
  Honiley 113, 116
  Hornchurch 90–95
  Isles of Scilly 57
  Leighton Buzzard 149
  Lindholme 129–131
  Medvejya Gora 69
  Ouston 113, 116
  Pembroke Dock 99–102
  Redcar 53
  Tangmere 116
  Upavon 14, 17
  Vendome 63
  Wilmslow 129
  Yatesbury 129–131
  41 Air School 112–113, 116
Royal Engineers 8–14
  Air Battalion 14
  Bechuanaland 8, 13, 139
  Boer War 8, 12, 13, 139
  China 12, 13, 139, 140
  School of Ballooning 8
  Suakin 11, 8, 9, 139
Royal Flying Corps
  Balloon Units 23–33
  Combat Reports 36
  Formation 14
  In WW1 18–49
  Medical Records 26–29
  Prisoners of War 46–49

Structure 30
Unit Records 30–36
War Diaries 30–32
Royal Naval Air Service
Armoured Cars 59–62
Armoured Trains 62
Admiralty Record 63
Casualties 64
Formation 16
In Royal Navy List 17
Ships 64
Splits from RFC 16
Squadrons 58
Russia (RAF in) 69–71

Salmond, Air Vice Marshal, J 76, 82, 112
Samson, Air Commodore, C R 16, 50, 53, 54
Service Records (Army)
Officers 8–9, 21–22, 125
Other ranks 9–12, 125
Service Records (Royal Flying Corps)
Service Records (Royal Navy) 51, 53
Service Records (RNAS)
Officers 51–56
Ratings 56–58
Service Records (RNVR) 54
Shanghai (RAF in) 86
603 Squadron 92, 94
605 Squadron 74, 109–110
Ships
HMS *Ajax* 119
HMS *Albion* 135
HMS *Ark Royal* 119, 120
HMS *Ben-My-Chree* 64
HMS *Eagle* 135
HMS *Glorious* 119
HMS *Hermes* 120
HMS *Nairana* 69
HMS *Pegasus* 70
Somaliland 71–72
Squadrons (RAF, RFC and RNAS)
1 Squadron 77, 79
1 (RNAS) 58
1 (Kite Balloon) 39, 40
2 Squadron 25, 31
3 Squadron 31, 40, 43
3 Squadron (RNAS) 58
4 Squadron 30, 40, 43
5 Squadron 31, 75, 79, 82, 83, 84
7 Squadron 68
8 Squadron (RNAS) 58
9 Squadron 39, 40, 43
11 Squadron 68, 79
12 Squadron 45, 68
13 Squadron 27
15 Squadron 43
22 Squadron 43
24 Squadron 43
27 Squadron 41, 798, 83, 84
28 Squadron 82–83
29 Squadron 46–49
31 Squadron 78, 82, 84
34 Squadron 22, 34, 35
41 Squadron 91, 92, 94
48 Squadron 68, 79
54 Squadron 91, 92, 94
60 Squadron 79, 83, 84
62 Squadron (OTU) 113
63 Squadron (OTU) 113
65 Squadron 91, 92, 93, 94
70 Squadron 68, 77
84 Squadron 68, 79
114 Squadron 78, 79
151 Squadron 113–116
264 92, 94
266 92, 94
600 Squadron 92, 94
Strubell, Major TFG 20
Strugnell, Group Capt W V 26
Suez Crisis 134–137

Thompson, Flt Lt A B 107
Tomaselli, AC2 P V 129–131
Trenchard, Marshal of the RAF Sir Hugh
15, 19, 64, 71, 76
Turner, AM1 AE 57
Turner, Flt Lt H J 107

War Office 12, 44, 135
Waziristan 83–85
Wheeler, WO Charles 101
Women's Auxiliary Air Force 117
Women's Royal Air Force 66–68
Formation 66
Officers 67
Other Ranks 67–68

Young, 2nd Lt J M 143–144